SuperMars

Other books by Ellis Silver

Humans Are Not From Earth

SuperMars

**Where We Really Came From
And Where We'll Go
When The World Ends**

Ellis Silver, PhD

ideas4writers

Copyright © Ellis Silver 2021

Published in Great Britain in 2021

by

ideas4writers
19 Crow Green
Cullompton
Devon
EX15 1EW

ideas4writers.com

ISBN 978-1-8383690-0-2

The right of Ellis Silver to be identified as the author of this work has been asserted by him in accordance with the Copyright, Designs and Patents Act 1988.

All rights reserved.

The contents of this work may not be stored, copied, transmitted, sold, or reproduced in any form without the permission of the publisher.

Cover design and layout by Dave Haslett

Cover images courtesy of
ESO/M. Kornmesser/N. Risinger (skysurvey.org)
NASA/JPL-Caltech
Pexels from Pixabay

Contents

1. Introduction..7
2. We Are More Martian Than Earthling.................................23
3. Why Mars Suits Us Better Than The Earth....................47
4. Why We Had To Leave Our Home Planet...................127
5. Mars Past And Present...153
6. Earth, Mars, Eden: The Differences And Similarities..173
7. Eden, Our Original Home Planet..................................187
8. Life On Mars...203
9. Making Mars Habitable..221
10. Humans On Earth..313
11. Our Future On Mars..355
12. Our Future In Space..369
13. Our Ultimate Future And Fate......................................383
14. References..391
15. About the Author..427
16. Index..429

1
Introduction

In the previous book, *Humans Are Not From Earth*, we looked at the evidence that proves we could not have evolved on Earth, and concluded we must have been brought here from somewhere else – another planet in another solar system.

We clearly have no place being here: we don't fit naturally into the Earth's environment, it makes us ill, and we're destroying the place. The timeline of human evolution on Earth is broken, full of gaps, and highly misleading. There are missing links that will never be found because they don't exist in that version of the model.

Most mainstream scientists cling to the Out of Africa theory, even though it's been repeatedly disproved. And they cling to the theory that modern humans are no more than 120,000 to 200,000 years old, even though we've found validated evidence, including DNA, that proves we've been on Earth for at least 400,000 years.

Our true origins obviously go back even further than that – not on this planet, but on our home planet, Eden*.

> *Eden seems as good a name as any for our home planet. When I suggested it in *Humans Are Not From Earth*, no one objected, so I'm sticking with it. The name comes from the Biblical story of how Adam and Eve (the first modern humans on Earth) were banished from the Garden of Eden. The Garden of Eden may have been a reference to our home planet. And the place they were banished to – which the Bible says contained hominins but no other modern humans – may have been the Earth.

Misguided mainstream scientists base their theories and hypotheses on the "definite fact" that we evolved on Earth. But that's not the case, and, as a result, their theories and hypotheses fall apart under the slightest scrutiny. They have to be revised repeatedly – and sometimes rewritten completely – every time new, indisputable evidence is discovered and accepted. But those major revisions and rewrites tend not to happen. They're quietly and conveniently "forgotten." Consequently, most mainstream scientists cling to their old theories. And the same old theories continue to be taught in schools, colleges and universities, as well as in broadcasts aimed at the general public, even though they're wrong and should have been abandoned years ago.

The gatekeepers of science would rather disregard the evidence in front of their faces than admit their original theories could have been wrong. And, of course, they would never (never!) accept that we might have evolved on another planet. Even though we did.

If you re-evaluate their theories and hypotheses with this one fact in mind – that we evolved somewhere else –

everything makes a lot more sense. Our true home planet is not the Earth. It isn't in this solar system. And it might not even be in this galaxy.

Exactly why we're here, in this region of space, is open to question. Some of the more open-minded scientists have proposed plenty of theories, and some of them have put forward a significant amount of evidence. But they're shunned by the mainstream scientific community, and their theories are branded "pseudoscience." The evidence is either taken away and locked up so no one can look at it, or it's dismissed as irrelevant, fraudulent or "impossible" and goes unexamined.

> As we saw in *Humans Are Not From Earth*, many mainstream scientists refuse to get involved with these theories, or examine the evidence, because they believe it could damage their careers. They stick with the "approved" theories, even when they don't make any sense.

It's interesting to note that our galaxy, the Milky Way, is located near the center of the KBC void – the largest known void in the observable universe. The void is two billion light years across, and all but empty. That makes it the perfect place to put us if someone wanted us out of the way. It means we're extremely remote from the rest of the universe, and effectively cut off from it. Furthermore, the solar system we live in is well away from the center of the Milky Way (also known as the "galactic center") where other civilizations are most likely to be located. Again, it's the perfect place to put us if someone wanted us out of the way.

People often say our radio transmissions and television broadcasts must have radiated out into space, and any extraterrestrial civilizations within a hundred light years of us could have detected them. But there probably *aren't* any civilizations that close to us. Most of them will be nearer the galactic center, which is 25,000 light years away. So our broadcasts won't reach them for another 24,900 years.

> A recent study estimated there could be 36 active civilizations in the Milky Way galaxy. But they are an average of 17,000 light years from each other. Conventional communications between them would be impossible – and our broadcasts won't reach our nearest neighbors for another 16,900 years.

Of course, we don't know how many civilizations there really are at the center of the galaxy. There could be millions of them. And they might all be communicating – or even trading – with each other. The big question, of course, is why are they there when we're all the way out here on our own? Is it a massive coincidence? Or did someone plan it that way?

We'll take a closer look at this later in the book.

Many readers of *Humans Are Not From Earth* agreed that the evidence suggests we did not evolve on Earth. But they pointed out that it correlates extremely well with the planet Mars. Many of the factors that damage us on Earth would benefit us there. For example, Mars's lower gravity, lower level of light, and the longer length of its days would suit us far better than they do here. If Mars was habitable, it would be a much better home for us than Earth.

Mars isn't habitable. Or at least it isn't habitable *now*. Until recently, most scientists believed it had *never* been habitable. But data returned by the USA's Martian space probes and rovers has proved otherwise. In fact, there's a strong possibility that Mars could have been habitable for billions of years, and it might only have become uninhabitable in the last few million years.

I'm certainly not claiming that humans *evolved* on Mars. We definitely did not.

And there have never been any crawling or flying insects, as one misguided scientist claimed. Life on Mars could never have evolved that far. In any case, what would it eat?

> The scientist said the insects' wings he had spotted in photographs of Mars were similar in size to insects on Earth. Many people have said that flight would be impossible in that case, because the Martian atmosphere is a hundred times thinner than the Earth's. In fact, flight *would* be possible (in theory), but the Martian insects would have to flap their wings a hundred times faster than terrestrial ones. Most flies flap their wings around 200 to 250 times per second, so Martian flies would have to flap them around 20,000 times per second, or 1.2 million times per minute. This is many times faster than neurons can fire, and it would take a *huge* amount of energy. So it's reasonable to conclude there are no flying insects on Mars.

Although we didn't come from Mars, it's clear that we must have come from a Mars-like planet. Eden is probably slightly larger than the planet Mars we're familiar with – a super-Mars if you like – but it would not be as large as the Earth. We'll look at the evidence for this later in the book.

> Many astronomers are looking for super-Earths, perhaps with the idea that they're a better place to live than the Earth[1-1], that they might become our future homes, or that a species like us might be living on them. But we would *hate* to live on a super-Earth; it wouldn't suit us at all. The astronomers should be looking for super-Marses – just like planet Eden we originally came from.

Until 2020, most astronomers believed Mars had once been covered in flowing rivers[1-2] which carved out valleys and river beds similar to those we see on Earth. Many people envisioned historic Mars as a lush green paradise. But recent evidence[1-3] suggests that Mars was completely covered by a thick ice sheet for billions of years. No land plants and no higher forms of life could have lived there. The likelihood is that the valleys and river beds were created by glacial action and by water flowing underneath the ice sheet. But that doesn't mean there will *never* be land plants and higher forms of life on Mars – as we'll see later in the book.

Today, Mars is a dry, deeply frozen desert with hardly any atmosphere and no magnetosphere to shield it from the solar wind, solar flares, and solar and cosmic radiation. We might not like living on Earth, but life on Mars today – and for the foreseeable future – would be horrendous.

That could change one day. In fact, it will *have* to, because our time on Earth is coming to an end. The only way our species will be able to survive the next few billion years is if we terraform* Mars and make it our new home.

> *Terraforming means reshaping a planet to make it habitable to life from Earth. It involves making changes to its magnetosphere, atmosphere, temperature, and ecology. The process is currently hypothetical. We believe it may be possible, but none of the steps have been attempted in real life, and most of them are way beyond our current level of technology.

> Surprisingly, Mars's wispy thin atmosphere provides a considerable amount of protection against radiation[1-4]. Some parts of the planet are just as safe to live on as the *International Space Station* (ISS). The ISS receives a much higher dose of radiation than the Earth does, but researchers believe it falls within safe and acceptable levels. However, the level of protection varies across the surface. Some regions receive such high doses of radiation that it would be hazardous to live there without shielding[1-5]. On the other hand, some areas such as Hellas Planitia, a deep-impact basin, receive very low doses. They would make ideal locations for the first Martian bases[1-6].

> The radiation shielding could be man-made, or it could be natural — in the form of underground tunnels, caves and lava tubes[1-7]. In some places though, even that wouldn't be enough.

> The underground habitats would need to be several miles deep, or heavily lined with man-made shielding to provide enough protection.

> Radiation shielding is generally made from stainless steel or other materials that would be expensive to ship from Earth. But researchers have discovered radiation-absorbing fungus growing at the destroyed Chernobyl nuclear reactor in Ukraine[1-8]. They've tested the fungus on the *International Space Station*, and believe it could be used on spacecraft, and grown on Mars to shield bases.

It's reasonable to conclude that life on Mars *could* have once existed. It might even exist there today, in a limited way[1-9].

Mars formed at about the same time as the Earth and Venus, and its surface cooled at about the same time too – probably a little faster in fact, as it's a smaller planet. Life appeared on Earth within a few hundred million years, and there's no reason why it couldn't have appeared on Mars as well. Conditions on both planets would have been similar, with large oceans and lakes[1-10], dense atmospheres, and comparable seasonal variations. Mars would have been significantly cooler, of course, as it's further from the Sun. The surface might have been covered in ice, just as Antarctica is on Earth, but the oceans beneath the ice would have been warmed by the molten core and mantel, and they could have been habitable.

> When the planets first formed, they would have been constantly bombarded with rocky debris. This would have kept them molten, at a temperature of around 3600°F (1980°C)

However, things quickly went catastrophically wrong on Mars. Something enormous collided with it around four billion years ago, and any life that existed would have been wiped out. Many people believe Mars lost its atmosphere and most of its water as a result of the tremendous impact. But that isn't what happened. The atmosphere took billions of years to leak away, and recent studies have found that most of the water is still present. Although there's no longer any water on the surface, most of it has moved inside the planet and it has been absorbed by the highly porous basalt rock.

There's no reason why life couldn't have re-established itself after the impact, even if it took hundreds of millions of years to recover.

The main reason why Mars lost its atmosphere is simply because it's smaller than the Earth, and its outer core cooled and solidified faster. As a result, the dynamo effect that generated Mars's magnetic field ceased. Magnetic fields are important because they shield planets from solar flares, solar wind, solar radiation and cosmic radiation. Having lost its magnetic shield, Mars was exposed to the full effects of these, and its atmosphere was gradually eroded.

> Mars still has a *small* magnetic field. It isn't strong enough to shield the planet from the solar wind and radiation, but we can detect it with our scientific instruments[1-11].

But it's important to understand that this happened *slowly* – the process took two to three billion years.

It's also important to understand that Mars still had oceans throughout this period – the water didn't disappear underground immediately. They gradually became smaller, over billions of years, but they may have remained habitable to lower organisms.

Mars's atmosphere wasn't breathable, of course, but nor was the Earth's at that time, and the Earth's oceans were full of life. Even as recently as one billion years ago (and perhaps more recently than that), there would have been enough air pressure on Mars to sustain primitive, aquatic, anaerobic life.

> The Earth's atmosphere started to become breathable about 2.45 billion years ago when the Great Oxygenation Event began[1-12]. Plants evolved and flourished, absorbed vast quantities of carbon dioxide, and began producing oxygen. There were two unfortunate side-effects of this. First, the oxygen was toxic to the anaerobic organisms that lived on the Earth at that time, and it caused the majority of them to become extinct. Second, the carbon dioxide in the atmosphere had been acting as an insulating layer to keep the Earth's surface warm. The plants absorbed so much of it that the surface cooled down – and it might even have frozen over for several million years. On the plus side, the Earth's oxygenation allowed higher forms of life to evolve. We'll look at this in more detail later.

The extraterrestrials (or "aliens" for want of a better word) who brought us here may have discovered Mars around this time and thought it would make an ideal new home for us. Its oceans may have been teeming with microbial life by then, just like the Earth's oceans were. The aliens might have begun terraforming Mars: making its atmosphere denser to cause global warming, raising the surface temperature and melting the ice, then seeding the surface and oceans with life, with the aim of turning it into a suitable place for us to live.

Their efforts would fail, of course, but they might not have realized this at the time. They may have persevered with their efforts for hundreds of thousands of years.

Their plan would have been to cover Mars in vegetation – much of it transplanted from our home planet Eden. The vegetation would have produced oxygen and made the atmosphere breathable. It would also have provided us with food to eat – and as it came from our home planet, it would have suited our physiologies and tastes, and we would have been able to recognize it.

The aliens might have planned to introduce animals, including flying insects to pollinate the plants and burrowing insects and worms to convert waste material and rotting vegetation into soil.

They might have then planned to ship us from Eden to Mars in enormous spacecraft.

But as we know, the Martian environment collapsed. So much of the atmosphere was eventually stripped away and so much of the surface water disappeared that the planet could no longer support us – or any other higher forms of life. Although this process took billions of years, the aliens would have been able to see it happening. But they might not have been able to stop it.

What the heck were they going to do with us now? They still needed a new home for us, after all – for reasons we'll look at later.

Well, as it happened, there was another planet not too far away. It wouldn't be the paradise they'd planned for us, but we should (just about) be able to survive there. It had plenty of vegetation too. It wouldn't be what we were used to, because plants from Eden wouldn't grow there, but at least some of it would be edible. The atmosphere wasn't quite right either, and the light and gravity levels were far too high, and there were all sorts of other issues. But they aliens had committed enough time and resources to the project by then – and they were probably sick of us. There would have been political and financial pressure to come up with a quick solution. So they switched to Plan B and dumped us here on Earth.

Unfortunately, it turned out that Planet Earth *really* didn't suit us at all. In fact, the first few groups of modern humans they brought here swiftly died out. The Earth would *not* turn out to be the quick solution the aliens may have been hoping for. But there were no other options.

After making some modifications to our DNA, and borrowing genes from some of the Earth's native hominins (once they evolved), they overcame the most serious problems. They dropped several more groups of us at various locations around the world. And most of them survived … for a few thousand years. And then they died.

Planet Earth was pretty hostile to modern humans: there were all sorts of natural disasters we weren't designed to cope with; the levels of sunlight and gravity were far higher than we were used to or could cope with; and the food, water and overall environment reduced our lifespans and our fertility.

In the end, only a few hundred members of the African group survived, along with a few tiny clusters elsewhere. This probably explains the origins of the mainstream scientists' Out of Africa theory. But it also explains why we can find evidence of modern human civilizations – including artifacts, teeth, bone fragments, and DNA – in several other parts of the world, including Australia and China, with much of it predating the African evidence.

Humans are an inherently violent and territorial species. But even though they could alter our genomes, the aliens retained those elements. Perhaps our violence and territorialism are so ingrained that they couldn't remove them. But I think that's unlikely; they could have made us docile and compliant if they'd wanted to. They must have retained those elements for a specific purpose. But you can be sure it was for their benefit rather than ours.

They must also have been aware of what we would do to the Earth's native hominins. We know they borrowed some of their genes and spliced them into us, so they obviously knew the hominins existed. Whether we out-competed the hominins for food and resources, or gave them diseases they had no immunity to, or actively killed them is subject to debate, but one way or another *we* caused their extinction.

Why the aliens adopted Plan B and brought us to Earth, rather than leaving us on Eden or ejecting us into space, is another one of life's great unknowns. Again, we'll look at some possible reasons later in the book. But we should be eternally grateful that they did so – even though they committed indirect genocide by causing the extinction of the Neanderthals, Denisovans and the other hominin races.

We must also be grateful that the aliens made considerable efforts to ensure our survival. After discovering we couldn't survive on Earth in our natural form, they went to a great deal of trouble to modify our genomes, splicing parts of the hominins' genomes into ours, and perhaps also helping (or forcing) us to interbreed with them. In doing so, they not only gave us the ability to survive on Earth, they also created a new species: modern humans.

It's interesting to think about what we might have been like in our natural form, before we became modern humans. When we lived on Eden, we lacked the hominin genes from Earth. We would have been *humanoid*, but we wouldn't have been human. We might have had enormous heads and looked more like "aliens." We might have been taller, covered in hair and looked more like the fabled Yeti or Sasquatch. Or we might have looked like something else.

The rest, as they say, is history – but not the kind of history you were taught in school.

In the first part of this book, we'll reconsider the evidence from *Humans Are Not From Earth* and see how different things would be if we lived on a Mars-like planet. We'll take into account the physiological evidence in our bodies and our DNA, and the geological evidence we can see on the Earth and Mars. We'll also consider what life might have been like, and what *we* might have been like, if we'd been taken to Mars rather than the Earth – assuming Mars was habitable then and it had remained habitable to this day.

Later in the book, we'll consider a workable way of terraforming Mars, not only making it fully habitable but turning it into a facsimile of our original home planet, Eden. We have a fantastic opportunity to adapt Mars to suit our physiologies in ways that the Earth doesn't. We'll also consider our possible futures on Earth, Mars, and as a space-faring race in the greater universe.

Whatever happens, you can be sure that the aliens who brought us here (and other alien races) are monitoring our every move. They've prevented us from destroying ourselves – and the planet – with nuclear weapons. But they might also prevent us from venturing further than our own backyard.

2

We Are More Martian Than Earthling

Although we didn't *come* from Mars, it's possible that life on Earth was *seeded* by Mars[2-1][2-2].

How could that have happened when both planets formed at around the same time? Surely life would have evolved on both planets at around the same time too?

It all comes down to the fact that Mars is significantly smaller than the Earth, which means its surface would have cooled faster after it formed.

> Mars's rapid cooling also caused its downfall. When its inner mantel and core cooled and solidified, it lost its magnetic field, which had shielded it from solar wind and radiation.
>
> The Earth's core is slowly cooling too, and the same fate undoubtedly awaits it in several billion years' time. But there's a good chance that the Sun will have turned into a red giant star before that happens. Even if it doesn't consume the Earth (which is a possibility) the Earth's surface will

> become too hot to sustain life. So when the magnetic field disappears, we won't be here to witness it. Hopefully, we'll have colonized other worlds – including Mars – by then, and evacuated the entire population to a safer place.

We've found evidence that primitive life-forms existed on Earth just 110 million years after the planet formed. That's a remarkably short timescale. Some people (including me) might call it *suspiciously* short. But let's assume, unlikely as it is, that those life-forms actually evolved on Earth. The same thing could have happened on Mars, but fifty million years earlier, because its surface reached a habitable temperature that much sooner[2-3].

Some of the primitive life-forms that evolved on Mars could have then traveled to Earth in fissures in rocks that were ejected into space when Mars was struck by meteorites.

> Meteorite strikes would have been a common occurrence in those early days. Millions of rocks and comets drifted around the solar system and were gradually mopped up by the planets as they finished forming. As they struck each planet, at thousands of miles per hour, they would have sent tons of rock blasting into space in all directions. Some of the (potentially) life-bearing rock from Mars would have landed on other planets, including the Earth.

> We've found dozens of rocks on Earth that originally came from Mars.

> The same process would have happened in reverse too. Life-bearing rocks from Earth would have been ejected into space when *it* was struck by meteorites. Some of them will have landed on Mars and the other planets and moons in the solar system. Some will have left the solar system, and they may have reached planets orbiting other stars. Those stars and planets might be many light years away, but if the rocks left Earth three or four billion years ago, they've had plenty of time to get there.

If any life-bearing rocks from Mars hit the Earth about 110 million years after it formed – or, in other words, just as its surface reached a habitable temperature – they could have seeded it with life. The primitive organisms inside those rocks could be the origin of all life-forms on Earth today. So, even if we *did* evolve on Earth, we could be Martian in origin.

> Humans did not evolve on Earth.

However, while life-bearing rocks undoubtedly hit the Earth and seeded it with life, it's unlikely that they came from Mars. I don't believe life could have evolved on Mars that quickly, nor could it have evolved on Earth as quickly as it appears to have done. The timescales are far too short. My hypothesis is that life on Earth (and perhaps Mars too) must have been seeded from a more ancient world outside the solar system.

> This process is known as panspermia.

The life-bearing rocks that hit the Earth 110 million years ago and seeded it with life, probably hit Mars too. But if any of the organisms survived, they would have remained primitive, anaerobic, and microscopic.

It's possible that life still exists on Mars today, in exactly the same form, but deep underground. But it wouldn't be visible to the naked eye.

> The closest landscape to Mars that we have on Earth is the Atacama Desert in Chile. A team of researchers from Cornell University in the USA have found a wide range of microbes in wet clay eleven inches below the arid surface. They suggest we might find similar microbes in Mars's Gale crater, which has similar clay-rich deposits[2-4].

The evidence (which we looked at in the previous book) suggests that humans didn't evolve on either Earth *or* Mars. It's more likely that we came from a different planet.

Our original home planet (which we're calling Eden in this series of books) is a small, Mars-like planet that isn't in this solar system and might not be in this galaxy.

We're ill-suited to life on Earth – which would be bizarre if we had evolved here. After all, we are (in theory) the most advanced species on the planet. How can we *still* not have adapted to the environment after four billion years of evolution, when every other species has? Why are we so different from every other species? And how can so many everyday things make us chronically ill?

> These chronic illnesses and disorders are not modern phenomena, or consequences of our modern lifestyles, as many people believe. If we look at ancient human skeletons and remains, we can see that our (modern human) ancestors had exactly the same issues tens of thousands of years ago, when they were simple hunter-gatherers. On the other hand, if we look at the remains of Earth's native hominins, such as the Neanderthals and Denisovans, we can see that they had no such issues. They were *perfectly* adapted to life on Earth.

> Although they're often called our *more primitive ancestors*, modern studies of the Neanderthals' physiologies and remains have found that not only were they not our ancestors, they also weren't as primitive as we originally thought. In fact, they may have been more sophisticated than us when we first arrived on Earth.

However, if we think about the long list of problems we have on Earth and compare them with what we know about Mars, things start to make more sense. If we lived on a Mars-like planet (not necessarily Mars itself, but a similar planet) we wouldn't have *nearly* as many issues.

We'll take a look at the full list of issues in a moment, but here are just a few examples:

- The lower level of sunlight on Mars (or a Mars-like planet) would suit us better. We wouldn't be dazzled, suffer from skin cancer, premature aging, or cataracts.

- The lower level of gravity on Mars would be better for us too. We wouldn't suffer from bad backs, we'd be able to move around more freely, without the weight of the world pressing down on us, and we wouldn't hurt ourselves as much if we fell over.

- The longer Martian day matches our internal body clocks more closely than an Earth day does. If we lived on Mars, we would feel that the days were long enough, we'd feel less tired, and we wouldn't need to rush to get everything done in time. We would no longer have the feeling that there are never enough hours in the day.

These three factors alone would take care of a significant number of our physical and mental health issues. Many of them would simply disappear, and we would feel far happier as a result.

In *Humans Are Not From Earth* we examined more than fifty factors – and a ton of supporting evidence – that prove we could not have evolved on Earth. But this raised two important questions: why are we here, and where did we really come from?

Where we really came from – the location of our original home planet Eden – is open to debate. Unless some kind-hearted alien visitors tell us, we might never find out.

> Some people say alien visitors *have* told them. But as the aliens seem to have given different answers to different people, or they gave implausible answers, we can't consider their evidence scientifically valid at the moment.

In *Humans Are Not From Earth*, I made the reasonable assumption that Eden might be a planet orbiting a nearby star, such as Alpha Centauri B (also known as Toliman), or another star no more than forty light years away. That would mean that Eden is in the same region of the Milky Way galaxy as we are.

But we now know that the Milky Way is near the center of the KBC void – the largest-known void in the observable universe. That's unlikely to be a coincidence, and it suggests we were placed here deliberately. If that's the case, it's likely that we were brought here from somewhere much further away. So Eden might lie in another galaxy, and perhaps in a completely different part of the universe.

That doesn't mean it took us billions of years to get here though. There are (theoretically at least) several ways of traveling from one part of the universe to another, faster than light speed, without breaking any of the known laws of physics. Examples include the use of warp drives and wormholes. The aliens who brought us here might have been able to travel the tremendous distance from Eden to the Earth in just a few hours or a few days. I'd imagine they then put restrictions in place to prevent other alien races from bridging the gap so quickly – or from reaching us at all. They would also have banned other civilizations from communicating with us. Or perhaps they put some sort of mechanism in place to prevent them from doing so.

> We know of numerous other alien races, and we've been visited by several of them. So the restrictions might not have been as effective as they'd hoped.

We also need to think about the other main question: why are we here? We'll consider several potential reasons later in the book. One possibility is that we either destroyed Eden or came so close to doing so that we had to be evacuated (or rescued) from it.

We've come perilously close to destroying the Earth a few times too – along with ourselves, of course. The closest was probably the nuclear era of the Cold War, particularly the Cuban Missile Crisis of 1962. But we've also caused several long-term environmental catastrophes, including climate change and the pollution of our oceans with plastics and chemicals. We've also caused a potential environmental disaster in the region of space surrounding the Earth. It's now littered with millions of pieces of debris. You can easily imagine the damage we would cause on other planets – and what we might do to any alien species that lived there.

In fact, there's no need to imagine this, because we have past form. We would seize their territory and resources, wage war on them, drive them to extinction, and so on. We did it to the Neanderthals and the other hominin species on Earth not long after we arrived. We did it to the indigenous people of America, Australia, New Zealand, and other places when we discovered them. And we would do it again if we came across any other inhabited planets – or alien spacecraft.

Our innate desire for destruction (and self-destruction) combined with our inherent violence and territorialism, makes us a *very* dangerous species to have on your doorstep. No wonder the aliens removed us to a place so remote from the other civilizations that we wouldn't even know they existed, let alone pose any kind of threat to them.

As an additional guarantee of their safety, it's likely that they also erased our memories. And they might have blocked some of our more advanced functions, such as our ability to think analytically. As a result, we have no memory of Eden, the way we used to live, the technology (and weapons) we used to have, the aliens who brought us here, or anything else.

When we arrived on Earth, we could forage and hunt for food and just about survive, but not much else. We were reduced to the same level as the Neanderthals and the other hominin species that were already here. The aliens may have hoped we would integrate with them. But that's not how humans behave.

We lived that way for at least 100,000 years – and perhaps much longer, as we've found evidence that modern humans have been on Earth for at least 400,000 years.

We might *still* have been living that way today, but something changed around twelve thousand years ago. Perhaps some of our memories returned. Perhaps the block on our advanced thinking capabilities was lifted. Perhaps we were able to pass information from one generation to the next via our genetic memories, rather than the (impaired) memories in our brains. Or perhaps we had some outside help from another alien race. After 100,000 to 400,000 years of living primitively, we suddenly developed agriculture, mathematics, architecture, industrial processes, written language, and the entire modern world, completely out of the blue and all in the space of a few centuries.

It seems unbelievable and inconceivable that we spent so long living primitively and then transformed ourselves into nuclear-powered, space-faring, fast-food munching city dwellers in such a short amount of time.

I suspect the real answer is that we had outside help. Another alien race almost certainly discovered us and gave us a few nudges in the right direction to see what would happen. Little did they know what they were unleashing, or how quickly and how far we would advance.

> We'll look at this in more detail in the next book in the series.

Another theory is that some of the blocks the aliens placed on our memories and cognitive abilities might have expired. We can't remember how or where we used to live, or how we got here, but we can now think reasonably clearly, analyze data scientifically, communicate our findings, work in teams, record our knowledge, and put it to good use.

Having found a galaxy that was handily located in the KBC void, and having found a suitably remote part of that galaxy to deposit us in, the aliens located three potentially habitable planets that could become our new home: Venus, Earth and Mars. But which one should they choose?

It might seem obvious now, but it would have been less obvious back then. All three planets would have been remarkably similar, and there might have been little to choose between them. All three are rocky planets, and they're all (just about) in the habitable "Goldilocks" zone around the Sun. In their early days, they all had thick atmospheres, oceans of liquid water, massive seas, lakes and rivers – and perhaps even life.

The aliens may have decided against the Earth at first, because even though it was more-or-less habitable, it didn't resemble our original home planet Eden and the Mars-like conditions we were used to. We would have found it incredibly hard to survive here. Among other things, the bright sunlight would have damaged our eyes and skin; the higher gravity would have caused us all sorts of musculoskeletal problems; we would have felt heavy, cumbersome, and sluggish all the time; and we would have spent much of our lives in pain.

Another problem was that the Earth already had a thriving population of hominins, including the Neanderthals, Denisovans, and two or three other species. With our superior intelligence, speed, agility and dexterity – and our propensity for violence and destruction – we would have out-competed them for food and territory and driven them to extinction. We might also have carried diseases they had no immunity to, and those could have wiped them out too[2-5].

Venus may have been habitable at that time too[2-6][2-7][2-8]. Today, it's scorchingly hot – the result of runaway global warming. Its average surface temperature is 864°F (462°C) and it rains deadly sulfuric acid, making it inhospitable to all known forms of life. But it wasn't always like that, and it may have been inhabitable until comparatively recently. For example, we've found evidence of shallow oceans that may have lasted for up to three billion years. We don't yet know if it ever had any vegetation, or enough fresh water to sustain a population of hominins, but it could have done. But two other factors ruled Venus out of the equation.

First, a day on Venus lasts longer than a year. A day (a single rotation of the planet) lasts 243 Earth days, whereas a year (a single orbit around the Sun) takes just 225 Earth days. As a result, one side of the planet spends months at a time in direct sunlight, with searing temperatures, while the other side is in freezing darkness. Even the hardiest microbes would find life difficult-to-impossible in these conditions. Higher forms of life would stand no chance whatsoever.

Second, a cataclysmic event around 700 million years ago destroyed any chance Venus had of becoming our future home. Something triggered a massive release of carbon dioxide from the planet's rocks. Under normal circumstances, the gas would have been reabsorbed and Venus would have eventually recovered. But in this case, something prevented that from happening. The most likely reason is that the release was triggered by massive volcanic activity. The resulting lava fields, which coated much of the planet's surface, may have formed a barrier that prevented the carbon dioxide from being reabsorbed.

> A similar thing happened on Earth at least once. It probably caused a mass extinction around five hundred million years ago. But in that case, the carbon dioxide was eventually reabsorbed.

The carbon dioxide remained in Venus's atmosphere – and it remains there today. It made the atmosphere denser, and a thick layer of sulfur clouds blanketed the entire planet, trapping solar radiation. The surface heated up, and the water evaporated and photodecomposed* into its constituent molecules, hydrogen and oxygen.

> *Photodecomposition is a chemical reaction in which compounds are broken down by photons. The process is also known as photodissociation or photolysis. Any photon with a high wavelength can cause this reaction, including visible light, ultraviolet light, x-rays and gamma rays.

The ozone in the Earth's atmosphere is *created* by photodecomposition[2-9]. Ultraviolet light strikes oxygen molecules made up of two oxygen atoms (O^2), and breaks them into separate atoms. They then combine with unbroken O^2 molecules to form ozone (O^3).

However, photodecomposition also causes the *destruction* of the ozone in the Earth's atmosphere. Ultraviolet light breaks man-made chloro-fluorocarbon (CFC) compounds into individual atoms of carbon, fluorine, and ozone-destroying chlorine free radicals.

Normally, new ozone is produced at more or less the same rate as it's destroyed, but during the CFC crisis this didn't happen. The rate of destruction increased substantially, leading to a massive hole appearing in the ozone layer, especially over the southern polar region. The hole allowed high levels of ultraviolet radiation to reach the surface, leading to a spike in skin cancer cases. Fortunately, once the production of CFC was banned worldwide, the rate of ozone destruction fell and the hole was able to repair itself.

Small holes still appear in the Earth's ozone layer from time to time, but they are short-lived and generally repair themselves within a year or two.

So that leaves us with Mars. When the aliens first came across it, it might have seemed a promising candidate for our future home. Its geological conditions suit us much better than those on Earth, and probably resemble those on our original home planet Eden. There would have been an abundance of water[2-10], and much of the planet would have been climatically stable, with an atmospheric pressure around fifty percent higher than the Earth's.

We know there was once a massive ocean that covered half of the northern hemisphere, and perhaps as much as a third of the total surface area. The ocean was at least two-and-a-half miles (four kilometers) deep. There would also have been inland seas[2-11], great salt lakes[2-12], huge rivers[2-13], and outflows thousands of miles long. In total there would have been enough water to cover the entire surface of the planet to a depth of around 460 feet (140 meters).

The surface of Mars is bone dry today, but there's still a significant amount of water underground. It wasn't all lost to space as we used to believe [2-14]. It became locked up in the ice caps and frozen reservoirs, it was absorbed by the highly porous basalt rock, and it was trapped between grains of sand where it formed a frozen slurry. If we were able to melt it all, it would cover the entire surface to a depth of at least 115 feet (35 meters). So around one-quarter of the original water remains.

> As well as melting the ice, we would also need to pump the water to the surface, keep it in its liquid state, and prevent it from draining away again. We would also need to restore the magnetosphere and give the planet a reasonably dense atmosphere and air pressure so the water wouldn't just

> evaporate and be eroded by particles streaming from the Sun. We'll look at some potential ways of accomplishing these things in Chapter 9.

We don't know if Mars ever had any life, but there's a good chance it did. As we've seen, primitive life-forms appeared on Earth almost as soon as its surface had cooled enough to support it. There's no reason why life couldn't have appeared on Mars within that timescale too.

> How life first appeared on Earth (or Mars) is the subject of heated debate, and there are numerous competing theories. Perhaps life evolved naturally within a few million years because the chemical balance was just right. Perhaps it arrived from somewhere else, such as a comet, meteor or "alien seed." Perhaps it was created by an intelligent designer. Or perhaps something else happened. We'll probably never know.

Even if there was no life on Mars when the aliens discovered it, it may have had the right conditions to sustain it. It probably had a carbon dioxide-rich atmosphere, which would have been fine for plants and other anaerobic organisms. The carbon dioxide's insulating properties might also have kept the surface warm enough for most of the water to have remained liquid.

The aliens may have begun terraforming Mars to make it habitable. Or, if it was already habitable, they might have begun seeding it with life.

They probably aimed to turn Mars into the perfect habitat for us: a lush paradise with plentiful food and water – similar to Eden – where we could live happily and comfortably.

And, of course, there would have been no other hominin populations for us to compete with and drive to extinction.

But the aliens' efforts would fail. Mars is a small planet and its outer core cooled and became viscous soon after it formed. The dynamo effect that generated its magnetic field all but ceased, its magnetosphere disappeared, its atmosphere was eroded by streams of solar particles, its surface water was lost, and its surface was exposed to deadly levels of ionizing radiation, as well as energy bursts from coronal mass ejections (solar flares) that sterilized it every century or two.

In addition, shortly after Mars's surface cooled enough for it to support life, the planet was struck by a huge asteroid. This happened around 4.48 billion years ago – just 120 million years after Mars formed. The impact crater was enormous – as large as Europe, Asia and Australia combined. Any life on the planet would have been erased.

> If there was any life on Mars at that time, some of it may have traveled to Earth in rocks that were ejected into space when the asteroid struck. Those primitive life-forms *could* have seeded the Earth with life.

Although it took millions of years, Mars eventually recovered from the asteroid impact. Life may have evolved again, or it may have arrived on rocks from Earth or other planets, in comets, or in "seeds" that one or more alien races may have sent to this solar system.

The two cataclysmic events – the asteroid impact and the loss of its magnetic field – would eventually prove terminal. But on a more positive note, the asteroid impact caused significant volcanic activity, and the basalt rocks that formed as a result were around twenty-five percent more porous than usual. Researchers believe they absorbed most of the surface water, preventing it from evaporating and being swept away into space with the rest of the atmosphere.

> Some researchers believe Mars's magnetic field was destroyed by the asteroid impact – or by a series of impacts[2-15][2-16]. I don't subscribe to this theory myself, but it's possible they're right.

> Interestingly, some of the rocks on Mars show alternating bands of magnetism. This indicates that the polarity of the planet's magnetic field reversed several times before it disappeared.

> Similar alternating bands can be seen on Earth[2-17]. The last polarity reversal was 780,000 years ago. A complete reversal takes between 7,000 and 22,000 years, and many geologists believe another one may be imminent[2-18].

The loss of the atmosphere and surface water was gradual, and it took at least three billion years to reach its current state. That's more than long enough for life to have appeared there – perhaps for a second time. There's a tiny possibility that more advanced life-forms could have evolved. Potentially, there was even enough time for primitive *primates* to have evolved.

But it's highly unlikely that life on Mars ever evolved beyond simple microbes, because the surface would have been repeatedly sterilized by energy bursts from solar flares once the magnetosphere was lost.

Life may have survived beneath the surface, and it might still exist there today, but again it's most likely to be primitive and microbial.

Mars's atmospheric pressure eventually fell from 150 percent of the Earth's to less than one percent of the Earth's, and all of the surface water disappeared.

> Small streams of liquid water trickle across the surface from time to time, and we've seen small amounts of it at the bottom of craters. But they don't last for long.

If Mars's atmosphere and surface water had disappeared *immediately*, the planet would have been frozen, dry and lifeless for the past four billion years. But, thanks to its once-abundant water supply and high atmospheric pressure, it recovered from the asteroid impact within a few million years. It may have remained habitable (or *potentially* habitable) for over three billion years afterward.

Studies of coastal erosion have found that a large ocean continued to exist for at least a billion years after the impact. The ocean would have been connected to a vast network of underground lakes. And that network probably still exists today, although all but the deepest lakes will now be frozen.

> Until readers asked me to investigate it, and I researched it properly, I mistakenly believed that Mars had been dry, frozen and lifeless for the past four billion years. After all, it was "common knowledge." I believed this common knowledge so strongly that I rejected the readers' claims and

> didn't bother to investigate them. I *knew* it to be a "fact" – when it was actually nothing of the kind. Fortunately, my readers were persistent enough – and I was open-minded enough – that I decided to research the facts for myself. If I hadn't, this book wouldn't exist. The lesson here is to never trust your beliefs and the things you've read or been taught. It's important to research and evaluate the evidence for yourself.

The aliens' efforts to terraform Mars probably went smoothly at first. But Mars had a thick atmosphere back then, and plenty of water. The magnetic field would have been weakening as the various layers of the mantel and core cooled and became "sticky," but it was still strong enough to shield the planet from the solar wind and radiation. There would have been little sign of the terminal decline that was to come.

But when the aliens returned to Mars, perhaps a few thousand years after they began the terraforming it, they would have realized the awful truth: Mars was dying. It could never become our second home, as they may have envisioned – or at least not without spending a heck of a lot of time and money on fixing it. And they might not have had the time, money, or (more likely) the political will to do that.

So what were they going to do with us now?

The evidence suggests they brought us to Earth instead. It was the nearest planet capable of (just about) sustaining us. And we'd still be in one of the remotest parts of the KBC void, and well out reach of the other civilizations. However, bringing us to Earth was a desperate solution to a desperate problem. It didn't suit us nearly as well as Mars would have done, and the aliens must have known we'd wreak havoc on the native hominins, as well as potentially destroying the environment. But there was no other choice. It was either the Earth or our extinction. And (for reasons we'll look at later) they had agreed not to kill us off.

They did have a couple of tricks up their sleeves though. As we saw earlier, they could erase our memories so we would never remember the lives and technology we once enjoyed, or have the knowledge or ability to recreate it. They could also impair our large brains and dull our senses, reducing us to the same level as the other hominins that lived here. I believe they did both of those things, and we remained in that state for the best part of 400,000 years.

> I'm sure the aliens would have impaired our brains and memories if they'd taken us to Mars too. If we had retained our vast technological knowledge and brainpower, we could have become a dangerous threat to other civilizations – even from the depths of the KBC void.

Unfortunately, the aliens' early attempts to rehome us on Earth did not go well. We've found evidence (primitive tools, settlements and remains) which suggests that the aliens brought several small groups of modern humans to Earth over a million years ago. This was probably an experiment to see how well (or how badly) we would cope. They dropped groups in different parts of the world, including Africa, the Middle East, China, Australia, and elsewhere. But none of those early groups survived.

We don't know why they died, or exactly how long they lasted, but I doubt they would have been able to reproduce. So they probably died out within fifty years of arrival.

We can see from our DNA that the aliens then began modifying our genomes. They might also have tried interbreeding us with the native hominins to give us the genes we needed to survive here. But the strange scars and splices geneticists have reported seeing in our DNA suggest they inserted or manipulated genes directly.

Around four to five percent of our DNA can be traced to the Neanderthals, Denisovans, and other native hominins. The type and proportion of hominin DNA varies across different human populations. For example, Neanderthal DNA can be found in everyone of European descent, whereas Denisovan DNA can be found in Asian populations, but not in others.

A small amount of interbreeding may have occurred naturally, but it's doubtful this resulted in many pregnancies (or *any* pregnancies), as the species would have been biologically incompatible. Even if one of the females somehow became pregnant, the offspring would have been deformed or infertile.

However, a small number *may* have reached adulthood and managed to reproduce. This might have happened just once in fifty years, but some geneticists say it could explain the amount of hominin DNA we have in our genomes today.

Personally, I doubt this could have happened. The genetic differences between ourselves and the hominins would have been too great. It's far more likely that the aliens "borrowed" some of the hominins' genes and spliced them into us.

Alternatively, the aliens may have created us as a hybrid species by taking eggs and semen from multiple species and manipulating them in their laboratories. Those species may have included us (in our original form), the other native hominins on Earth, and perhaps the aliens themselves, or other alien species.

Eventually, after several further failed attempts to rehome us on Earth, and more genetic manipulation (and possible interbreeding), one or two of the groups survived. We don't know where these groups were based. East Africa is the most popular suggestion among anthropologists, as it ties in with the Out of Africa theory. But Australia and China are stronger contenders in my opinion.

Some of this group's members may have set out to explore the world. They established temporary settlements along the way (in India and the Middle East) before eventually settling in Africa.

This could explain the *"Into* Africa" theory that's becoming increasingly popular among anthropologists. It also explains why we can find modern human remains in China, Australia and the Middle East that are up to 100,000 years older than anything we've found in Africa.

The members of this group are our direct ancestors. In fact, all modern humans may have descended from a single male and female within that group. Perhaps we should call them Adam and Eve.

> A small group from that period might also have survived in Europe, as we've also found remains there that predate the African ones. It's possible that the migrating Australian/Asian group went northward into Europe after reaching the Middle East, and then migrated into Africa later. But I believe the early Europeans were most probably a separate group.

> Today's Asian and European populations have different amounts of Neanderthal and Denisovan DNA in their genomes. This is a further indication that those populations came from separate groups.

Summary

- We evolved on another planet in another solar system – and possibly in another galaxy in another part of the universe.

- We needed to be relocated – for reasons we'll consider later – and Mars was considered a good candidate. It was remarkably similar to Eden, and, most importantly, remote from other civilizations we might pose a threat to.

- The Martian environment collapsed before we could be relocated there, and we were brought to Earth instead. But this was a desperate measure, as we would struggle to survive, and we would pose a serious threat to the native hominins.

- The aliens manipulated our genomes to give us the facilities we needed to survive here. After multiple failed attempts, one reasonably large group (most likely from Asia and/or Australia) and a smaller group (from Europe) managed to gain a foothold and migrated to east Africa. The members of these groups are our direct ancestors.

In the next chapter we'll take a closer look at our physiologies and see how we're better suited to living on Mars than on Earth.

3
Why Mars Suits Us Better Than The Earth

In *Humans Are Not From Earth* we looked at more than fifty factors, and hundreds of pieces of evidence, that prove we could not have evolved on Earth. In this chapter we'll look at those same factors again, but this time we'll evaluate them based on the premise that we came from a Mars-like planet, and we may have been destined to live on Mars not Earth when we were brought to this solar system.

It's likely that our original home planet Eden closely resembled Mars, and we would feel perfectly at home on Mars if it was habitable. We feel out of place on Earth because we didn't originate here, we were never intended to live here, it bears little resemblance to our home world, and we arrived here far too recently to have adapted to our new environment.

As we look at Mars in this chapter, we'll focus on the *historic* Mars, as it was a few million years ago, rather than the way it is now. We'll assume that at least some of its surface is temperate, and that it retains enough of its atmosphere and surface water that we could live there comfortably. We'll also assume it has a plentiful supply of edible plant matter – which the aliens may have transplanted from Eden.

This is important because it means:

- Mars will have a stable, breathable atmosphere

- we'll have plenty of food, and it will be the type of food we're used to, that suits our bodies best, and that we like to eat

- we'll instinctively recognize which plants are edible and which ones are not

> The aliens probably tried transplanting the same plants on Earth, but found they didn't grow here. So we have to eat the Earth's native plants, which we don't instinctively recognize, don't like the taste of, and which don't suit us particularly well.

To get around this problem, we've selectively bred just about every edible plant species on Earth to change its taste, size, color, texture and nutritional composition.

> As we've developed new agricultural techniques, we've further altered each species to improve its yield, resistance to disease, and ease of harvest. We've also reduced its water and nutritional requirements, so it grows in different soil types and different environments. I'm sure we would have done exactly the same with the native species on Eden (and Mars) too.

Major evidence

Let's start with the main issues we have with the Sun here on Earth:

- **The Sun dazzles us**
- **The Sun damages our eyes**
- **The Sun damages our skin**
- **We don't seek shade as other animals do**

Mars is further away from the Sun and receives just forty-four percent of the solar radiation that reaches the Earth. As a result, it wouldn't damage our eyes or our skin, there would be no need to seek shade from it, and we'd find the level of light comfortable, even on the brightest days.

The same would also be true on Eden. This suggests that:

- Eden might be a similar distance from its star as Mars is from the Sun

- or the solar radiation emitted by Eden's star might be weaker than that emitted by the Sun – for example it might be a red dwarf

- or Eden's star might be smaller than the Sun

- or Eden's star might be larger and/or brighter than the Sun, but Eden might be much further away from it

- **We don't cope well with blue light**[3-1]

If you've ever worn a pair of blue-blocker eyeglasses, you'll have noticed that you immediately gain a significant improvement in the clarity of your vision. But it goes much further than that. There are medical conditions, such as dyskinesia, where tics, tremors, uncontrollable movements, and other problems immediately disappear when the sufferer wears blue-blocker eyeglasses[3-2].

This suggests we came from a planet with a significantly lower level of blue light. That might be because Eden's star is more red, yellow or orange than the Sun. For example, it might be a red dwarf, or it might be a K-type main-sequence star (the Sun is a G-type main-sequence star and is white). Alternatively, it might be because Eden's atmosphere blocks more blue light than the Earth's does. (This is certainly the case on Mars at the moment.) Or it might simply be because Eden is further away from its star than the Earth is from the Sun. Blue light has a lower wavelength than red light, and more of it gets deflected by the molecules it encounters during its journey, so less of it reaches the planet. Red light has a longer wavelength and continues unimpeded. (Again, this is also true of Mars.)

- **Our vitamin D levels are too low**

Initially, this might seem to be an anomaly. If our vitamin D levels are too low on Earth, then surely we would have an even bigger problem on Mars? Vitamin D is essential to our health in many different ways, and our bodies create most of what we need from cholesterol. The problem is that our bodies can only do this in the presence of bright sunlight. (Well, kind of …)

We can also obtain vitamin D from certain species of oily fish, but we'd need to eat an awful lot of it to meet our daily requirements. And, as we saw in the last book, we're biologically designed to eat plants not meat, so we shouldn't really eat fish anyway. So, what's going on here? Does this prove we *couldn't* have evolved on a Mars-like planet?

Well, no. This is one of those areas where the public's knowledge of science is slightly askew. While most people believe we need sunlight to convert cholesterol into vitamin D, what we *actually* need is UVB ultraviolet radiation. As Mars doesn't have a magnetic field to shield it, and its atmosphere is thinner than the Earth's, more than enough UVB reaches the surface to satisfy our needs. In fact, far too much UVB reaches the surface these days, and it would be lethal. But let's go back a few million years. There would still have been a weak magnetic field shielding the planet, and the atmosphere would have been reasonably dense, so the level of UVB would have been perfect.

> **OBJECTION: Mars would have needed a thick atmosphere in order for us to live there. That would have blocked most of the solar radiation, including the UVB.**
>
> Mars is significantly smaller than the Earth and its gravity is weaker, so its atmosphere should be naturally thinner anyway (in terms of its height from surface to space). In ancient times, Mars's atmosphere was fifty percent denser than the Earth's, but we're looking at it as it would have been a few million years ago, when a significant amount of it had been eroded. It would only need

> the right mixture of gases (mainly nitrogen and oxygen and a small amount of carbon dioxide) and just enough pressure that we could breathe it. The air pressure wouldn't need to be anywhere near as high as it is on Earth. In fact, we'd feel much more comfortable if it was fifty percent lower[3-3].

> People with breathing problems sometimes travel to mountain resorts to recover because they can breathe more easily there. The air pressure in these resorts is typically around half what it is at sea level. This also suggests that the air pressure at sea level on Eden is around half what it is on Earth.

> A good example of a high-altitude facility for the treatment of breathing conditions is the renowned Instituto Pio XII, located next to Lake Misurina in the Italian Dolomites[3-4].

Although you can't see it or feel it, the air on Earth is *really heavy*. There's about a ton of it pressing down on you right now – roughly the same weight as a small car. You don't notice it because you've grown used to it. But imagine how much better you'd feel if the car wasn't there.

As the air becomes denser, we find it more difficult to breathe. We feel heavier and more depressed as the weight of the world presses down on us. It gives us headaches and makes our ears and sinuses hurt. If it was *really* dense – as it is on planets such as Venus and Jupiter – it would crush us.

On the other hand, if we go too high, such as the summit of Mount Everest, the pressure is too low for us to be able to breathe properly. That's why most climbers carry personal oxygen supplies with them.

So it's reasonable to assume that Eden has a similar atmosphere to Mars (as it was several million years ago).

- **Loss/lack of body hair**
- **Wearing clothes**

This is an interesting anomaly. We can't live in most parts of the world without clothes. But why don't we have thick body hair, as the other primates do? In our original form – when we lived on Eden and before we acquired hominin DNA – we might have been covered in hair. As well as keeping us warm, it would have protected us against the high levels of UVB radiation on Eden.

> Some researchers argue that we lost our hair when we started wearing clothes, because we no longer needed it. They suggest that we started wearing clothes when we started exploring the colder regions of Earth. But that doesn't make sense. As we lost our hair, we would have needed thicker clothes or more layers to compensate. Why didn't we simply grow thicker hair as we migrated to the colder regions? Animals such as polar bears and arctic foxes survive perfectly well in freezing conditions with only their dense fur and body fat to keep them warm. We struggle to survive there even with multiple layers of clothes. Why aren't there any communities of hairy people living there?

3. Why Mars Suits Us Better Than The Earth

> There's another argument that we may have lost our hair because it was full of parasites. Again, that doesn't make sense because the other primates still have their hair, and they take care of the parasite problem by grooming each other. Humans enjoy grooming each other too, so I don't believe this can be the reason.

The aliens may have removed our body hair when they manipulated our DNA to enable us to survive here. They may have done this deliberately or accidentally. For example, the levels of UVB radiation might be so much lower on Earth than on Eden that we wouldn't be able to survive here if our body hair blocked it. Alternatively, the aliens might not have had any body hair, and when they replaced some of our genes with theirs, we may have lost our hair. Or they might have altered some of our genes to address a particular issue we were having – and to enable us to survive on Earth – but they didn't realize those genes also controlled our body hair.

Whatever happened, the result was that we lost most of our hair.

It's worth noting that the climate on Mars would have been cooler than it is on Earth. Eden's climate would almost certainly have been cooler than the Earth's too. So we probably had body hair when we lived on Eden. And, if we had been taken to Mars to live, we might have kept it.

> The first groups of humans that came to Earth may have died out because they had no body hair and they couldn't cope with the cold. The aliens may have introduced us to the concept of wearing clothes (or animal skins) to keep us alive.

Interestingly, our body hair might also have helped us get more vitamin D. In sheep, for example, ultraviolet radiation reacts with the lanolin in their greasy wool to create vitamin D, which the sheep absorb through their skin. Chemists use the same process to create vitamin D supplements. The same process would work in us if we had thick, greasy body hair.

> We might not smell very nice, but at least we'd be warm and we wouldn't suffer from rickets and other diseases associated with vitamin D deficiency.

On Earth, the level of solar radiation is fairly low, as we're protected by the magnetosphere, the ozone layer, cloud cover, and our clothes. Most of us also live outside the tropics and spend much of the day indoors. This is why so many of us suffer from vitamin D deficiencies or need to take supplements.

> On average, forty-two percent of Americans are deficient in vitamin D – and the problem is even worse in some populations. People who suffer from this deficiency develop chronic health problems that will become increasingly serious as they get older[3-5].

- **Our response to the seasons**

We don't cope well with the seasons on Earth. In summer, we overheat and millions of us suffer from hay fever. In winter, we get cold, breathing becomes more difficult, our mucus membranes dry out, we become more susceptible to disease, food can be scarce, our elderly and infirm can suffer from hypothermia, and many people suffer from a form of depression known as Seasonal Affective Disorder (SAD) because they don't get enough sunlight. The strange thing is that we can reproduce all year round, rather than at optimal times to ensure our offspring's survival. We are the only creatures on Earth that do this[3-6].

Hay fever

Let's deal with the hay fever issue first. This is an allergy to the type of pollen that's found on Earth, and it's a clear sign that we probably came from somewhere else.

> It's surprising that even though we've lived on Earth for at least 400,000 years now, we still haven't gotten used to the pollen.

> Having said that though, it's likely that most of our ancestors arrived on Earth more recently than that. As we'll see later (in the revised timeline of human evolution) the aliens brought us to Earth in seven large groups. The most recent group (the South Americans) only arrived around 100,000 years ago, when most of the earlier groups had died out (after a series of natural disasters), and we were an endangered species. So we haven't had as much time to get used to the Earth's pollen as we might have done. 100,000 years is quite a long time, but it's obviously not long enough for everyone to have evolved enough resistance to it.

We can assume that the plants on Eden must be different from those on Earth, and that the plants on Eden wouldn't cause us any problems.

If we consider Mars, there are three possible scenarios:

- Mars had no plant life of its own when the aliens first encountered it, and they seeded it with plants from Eden. We would have been accustomed to these plants, and would not have suffered any allergies there.

- Mars had *some* plant life, and the aliens supplemented it with plants from Eden. We might have suffered from allergies to the Martian plants, but not to the ones brought from Eden.

- Mars might (or might not) have had some plant life, but plants from Eden wouldn't grow there. The aliens may have seeded it with plants from somewhere else. We might have been allergic to its pollen, just as we're allergic to some of the pollen on Earth.

OBJECTION: Other animals, including some cats, dogs and cattle, suffer from hay fever too.

True. But that's because we've selectively bred or genetically manipulated certain types of plant. The pollen they're allergic to isn't the *natural* pollen they evolved to cope with. The pollen most animals are allergic to comes from *selectively bred* flowers, grass such as lawns, and grass-based crops such as cereals and grains.

> **OBJECTION: In that case, *we* must also be allergic to the manipulated pollen, not the naturally occurring types.**
>
> No. Many of us are allergic to the *natural* pollen on Earth too. The worst offender is naturally occurring grass pollen, as well as pollen from other types of grass, trees (some of which may have been genetically manipulated or selectively bred), and weeds (which definitely haven't been altered)[3-7].

Seasonal variations

Seasons on Mars are nearly twice as long as they are on Earth. That's because Mars has a similar axial tilt to the Earth, but takes twice as long to orbit the Sun. Apart from their length, the seasons on Mars are remarkably similar to those on Earth.

> We don't cope well with seasons, which suggests that Eden probably doesn't have them. That must mean it doesn't tilt on its axis. As a result, the climate and length of day will remain the same throughout the year. This might be one of the major differences between Eden and Mars.

Let's think about how we would cope with the longer seasons on Mars.

The longer summers should be good – as long as we don't suffer from hay fever. They wouldn't be as hot as they are on Earth, but as long as the atmosphere was reasonably dense, they should still be pleasantly warm and comfortable.

Spring and fall would be similarly long, but they don't generally cause us many health issues. Most people on Earth enjoy them, and I think they'd enjoy them on Mars too.

But winter would be a different issue. We don't cope well with the three-month winters on Earth. How would we cope if they lasted for nearly six months, and the days were even darker and colder than they are here? It sounds depressing, doesn't it? I don't know about you, but I think I'd want to hibernate.

And that might be exactly what would happen. As we saw in *Humans Are Not From Earth*, there's some evidence that we have an inherent ability to do it.

> I'm slightly torn over this issue, because if we truly have the ability to hibernate, then Eden *must* have seasonal variations. If it didn't have seasons then we wouldn't have evolved the ability to hibernate. However, our ability to hibernate has only been seen in a handful of cases, and it hasn't been scientifically tested. The cases we've seen were more like extended hypothermia than the true hibernation we see in other animals. We'll have to delay making a proper judgment until this issue has been fully tested. Unfortunately, testing it would be highly unethical. So I don't think it's going to happen any time soon.

We may have gained the desire (and perhaps also the ability) to hibernate as a result of alien experimentation. The aliens may have given us the hibernation genes so we could cope with the winters on Earth. But the hibernation genes might not have worked as well as they expected.

So many people may have died when they attempted to hibernate that they aliens may have disabled the genes. So we've retained our desire to hibernate, but we no longer have the ability to do so.

Another advantage of hibernation is that we would have had little or no contact with others during the winter months, when we're more susceptible to pathogens like cold and flu viruses.

> Although we think of flu as a winter disease, the virus that causes it is present all year round. The cold winter air dries out our mucous membranes and makes us more susceptible to it. So it's actually our *response* to flu that's seasonal, rather than the disease itself. If you've ever caught flu when it wasn't winter and you wondered how that could have happened, it's probably because your mucous membranes dried out. That can happen for any number of reasons, not just because the weather is cold.

- **Our response to the environment**

We don't enjoy the environmental conditions on Earth, and we constantly strive to make things better for ourselves. Often, this is to the detriment of the planet and the other species we share it with. Would we do this on another planet where the conditions suited us perfectly? The short answer is, of course, yes. We're a highly creative and innovative species and we can always spot an opportunity to "make things better" – even when they're already perfect. If we didn't do this, I imagine we'd get extremely bored.

The problem is that we wouldn't make things better in the long-term. We might make them better in the short term, but we would eventually realize how much damage we'd caused. And then we'd have to try and undo it – if it wasn't too late. We've seen this pattern time and again on Earth.

• Chronic illnesses and medical conditions

In *Humans Are Not From Earth* I presented a long list of chronic health issues and said that every single person on Earth suffered from at least one of them. Most people suffer from several of them. I still stand by that claim, as no one has provided any evidence to the contrary. I personally suffer from at least eleven of the issues on the list, yet most people would regard me as a picture of health.

Every one of the issues on the list can be traced back to the fact that we don't belong on Earth, and the conditions here don't suit us. The gravity is wrong, the sunlight is wrong, the climate is wrong, the seasons are wrong – and confusing, the days keep changing length and leave us feeling exhausted, the plants and food are wrong so we're overfed yet undernourished, the water is wrong, and so on. We wouldn't have suffered from any of these issues on Eden. We might not have suffered from them on Mars either, but it depends on how similar the two planets are, and how well the aliens succeeded in transforming Mars into a facsimile of Eden.

That's not to say we wouldn't have created more than a few issues of our own, though. For example, we would almost certainly have created terrible things called "jobs." Jobs lead to pressure and stress, depression, anxiety about money, constant striving for promotion, competitiveness and ill-feeling, disputes, unionization, and so on. So our modern lives on Mars might not have been so different from our modern lives on Earth. We might have enjoyed better health because of the more suitable climate, gravity, light level, length of days, lack of allergies, and so on. But we would also have had our full range of modern illnesses – the ones that are connected with our lifestyles, stress and anxiety. Our ancestors would not have suffered from any of these.

> The chronic issues I presented in the original list were all (except one) issues that our ancestors also suffered from. They were *not* caused by our jobs or lifestyles.

But there's another issue that many biologists believe is the root cause of most of our illnesses, and it goes back hundreds of thousands of years…

- **We eat meat (but we shouldn't)**

As we discussed in detail in the previous book, we are not omnivores as most people believe. We're herbivores that also eat meat. It's an important distinction. True omnivores have the capacity to digest and process meat properly, but we don't. Our stomach acid is too weak, and the meat stays in our guts for too long, decomposes, and releases toxins. Adopting a vegetarian or strict vegan diet, even right from birth, won't necessarily help because our ancestors ate meat too. This caused genetic disorders and weaknesses that they passed down to us.

There were major advantages to eating meat when we first arrived on Earth – and even before that. It's packed with much more protein than we could have gotten from the plants of that era. Our brains need huge amounts of energy to function properly, and meat provided it. In fact, many anthropologists believe eating meat led to a massive increase in our intelligence.

> There's another clue here about why our brains and memories might not have functioned properly when we first arrived on Earth: we may have lacked sufficient protein. We may have been used to high-protein diets on Eden, but the plants on Earth lacked the level of nutrition we needed. We may have started eating meat to regain our brain and memory functions – and we probably learned this from the native hominins on Earth, who were already meat-eaters.

Modern, selectively bred and genetically modified plants are highly nutritious and packed with protein, so a vegetarian or vegan diet *is* now feasible. But in ancient times, non-meat-eaters – which would have been *all of us* when we first arrived on Earth – would have lacked energy and brain power, would have been slow and weak, and probably would have died younger than the meat-eaters.

> This may have been one of the main reasons why the aliens believed we would integrate with the native hominins rather than drive them to extinction. As the hominins were meat-eaters, they would have been faster, more agile, and more intelligent than us. Once we became meat eaters too, it didn't take us long to catch up and then overtake them.

The evidence we've found suggests that the Earth's native hominins began eating meat around 2.6 million years ago[3-8]. When the first groups of modern humans arrived (as herbivores) around 400,000 years ago, and they encountered the hominins, they probably copied them and started eating meat too. The hominins might even have invited us to share their meals.

If we think about our home planet, Eden, we can assume that, based on our physiologies, our diet would have been entirely plant-based. It's also likely that the plants there were nutritious, full of protein, and naturally tasty – much more so than the ones on Earth.

> However, the plants on Eden would not have been as nutritious as the plants our scientists have developed on Earth in recent years. This would have kept our population size in check, and kept our weight down.

When they tried to make Mars more like Eden, the aliens may have planned to transplant many of the food crops we were used to. But when they were forced to bring us to Earth instead, they would have had another problem on their hands: the food crops from Eden wouldn't grow here.

They *may* have modified some of those crops and transplanted them here more recently – perhaps around 400,000 years ago when the first large groups of modern humans started surviving on Earth. The crops they brought from Eden *may* have become the basis for our modern fruit, vegetables, nuts, berries and cereals. But it's far more likely that *we* modified the Earth's native crops, and created the modern versions ourselves.

Now that we've developed highly nutritious crops here on Earth, and scarcity is no longer an issue, our population is expanding massively, both in number and in waist size. And that's another problem.

● We can't drink the water

We can't drink from impure water sources on Earth, such as puddles, ponds, lakes, rivers, muddy streams and rivulets, gutters, and so on. We'd be hospitalized in hours if we tried it. Yet other animals come to no harm whatsoever – as long as the water doesn't contain any *man-made* pollutants, of course.

There are three reasons for this, as far as I can tell:

- Eden must have an ample supply of naturally pure fresh water.

- Or the water on Eden must be sterile and contain no organisms that can harm us – although this seems unlikely.

- Or we developed a resistance to the waterborne organisms on Eden. Once we arrived on Earth, we had no resistance to the native waterborne organisms, and we have not yet had enough time to develop any resistance.

    Since every developed country now treats its drinking water with chemicals, and people in most developing countries boil their water before drinking it, it's unlikely that we'll ever develop any natural resistance to these organisms.

The most likely scenario is that Eden must have had an abundance of pure water from underground springs. It would have been filtered through rocks, and almost entirely free of microorganisms and dirt.

This might also have been the case on Mars. We know it once had a plentiful supply of fresh water, which filtered up through rocks to form natural springs. There would also have been plenty of pure, fresh water that melted from the ice caps. We would not have needed to drink the water in the lakes, rivers and streams. Here on Earth, although there's a plentiful supply of naturally filtered spring water and mineral water, it isn't always located near centers of population. It can, however, be bottled and distributed to anyone who needs it (and can afford it).

> We probably had an inherent resistance to the potentially harmful waterborne organisms in the lakes, rivers and streams on Eden – just like the other animals that lived there. We would not have had that resistance on Mars if it had its own native population of waterborne organisms.

> I assume the aliens would not have brought any harmful microorganisms to Earth or Mars from Eden.

- We dislike natural foods
- Food cravings
- We lack sufficient calcium in our diet
- We don't have a fixed diet
- Excessive growth of internal parasites

All of these factors confirm that the naturally occurring edible plants on Earth don't suit our needs or our tastes. On Eden there was an abundance of protein-rich flavorsome plants, but the stuff we found on Earth was poisonous, tasted horrible, tasted bland, or lacked nutrition.

After finding out (the hard way) which plants were poisonous or tasted horrible, we managed to survive on the remainder, supplementing it with cooked meat, for almost 400,000 years. But this would have been a *survival* diet. It kept us alive (for a while) and just about able to function, but it wasn't particularly nutritious or appetizing. Everyone died young; few people survived beyond the age of thirty.

This continued until around 12,000 years ago, when we woke from our 400,000-year "daze" and we were finally able to access our brains' full resources.

During our "intelligence revolution," we developed selective breeding, which enabled us to adapt the Earth's native crops to our taste, and increase their nutritional content. We also made them more resistant to disease, and increased their yields. But, humans being humans, we then took things too far, surpassing the taste, nutrition and availability of Eden's crops by a considerable margin. This has led to a population explosion, and an obesity crisis in the developed world.

Some crops that were present on Eden are missing on Earth. These crops would have been rich in calcium and other minerals. None of the crops on Earth contain such high levels of these minerals or satisfy our inherent need for them. As a result, we not only have cravings for mineral-rich substances, but many of us also suffer from mineral deficiencies, especially calcium. This further contributes to our chronic ill health.

The aliens almost certainly collected mineral-rich crops from Eden and attempted to grow them on Mars. Mars has rich deposits of calcium and other minerals, so it's probably very much like Eden in that respect. It's highly likely that the crops from Eden *would* have grown there. Here on Earth, the calcium and mineral levels are significantly lower (except for a few locations) and those crops would not have grown.

Our lack of a fixed diet can be explained by the fact that when we arrived on Earth, we were desperate to find the nutrition and minerals we needed. We were willing to try *anything*. As long as it didn't harm us (too much), we would eat it. A wide range of native plants proved edible, and appealed to different tastes. Some people happily ate certain plants, while other people found those same plants revolting.

This begs the question: why do we have different tastes? Researchers believe it depends on what we were exposed to in the womb, as children, as we grew up, and so on[3-9]. Certain families and cultures eat particular meals, knowing that their ingredients won't harm them. They eat the same types of meal over and over again, and they become engrained in their culture. Their children eat what their parents eat – they're usually given no choice in this – and they learn to like it too (or go without). Certain foods might be "an acquired taste," but, over time, and with repeated exposure, most people *do* acquire that taste.

Members of those cultures then meet people from other cultures who eat different types of food. They befriend them, and trust them to serve them food that won't harm them. They might discover that they like those meals too, and adopt them into their own culture. (Or they might dislike them and choose to avoid them in future.)

Alternatively, we might have a bad experience with one particular type of food (or drink) that most people enjoy. Perhaps we eat a rotten nut, for example, and it makes us ill. Subsequently, we can't *bear* to eat another nut, even if we're sure it's good.

On Eden, and probably also on Mars, our diet would have been more restricted – just as it is for every other animal. We would only have eaten a narrow range of plants, yet we would have gained all the nutrients we needed.

We would all have eaten the same narrow range of plants for hundreds of generations, and we would therefore have developed a strong taste for them. They would also have been the only plants that appealed to us. Everything else would have tasted or smelled foul, and we would have avoided it – even if it was actually good for us. We wouldn't have sought out different flavors or textures, or tried food from other cultures, because everyone would have eaten the same thing.

The issue of parasites growing excessively inside us also comes back to the fact that we didn't evolve on Earth. If we had, the parasites would have evolved alongside us, not interfered with our health to any great extent, and would have grown no larger than they do in other animals. We wouldn't even have known most of them were there.

This would also be the case if we still lived on Eden.

But here on Earth, our unfamiliar digestive systems provide the parasites with far more nutrition than they're used to – and much more than they're able to cope with.

As a result, they multiply, become obese, block our digestive tracts, and not only kill us but themselves too.

We don't know what the situation would have been on Mars. It depends on whether the food crops were native to Mars, transplanted from Eden, or brought in from somewhere else.

It also depends on whether those crops were nutritionally similar to those on Eden. If they were, then there wouldn't have been a problem. But if they were not, then we may have been faced with the same situation that we faced when we were brought to Earth. We were forced to live in "survival mode" for hundreds of thousands of years, and suffered from all sorts of chronic health problems as a result.

If we had lived on Mars, we might (or might not) have managed to break out of survival mode. If we didn't do it ourselves, the aliens might have given us a "nudge" in the right direction. And we might have eventually managed to selectively breed the crops into something tastier, more nutritious, and, ultimately, life-transforming.

- **Back problems**

From *Humans Are Not From Earth*:
Some scientists believe the Earth's gravity is twenty to forty percent too high for us. As a result, we feel heavy, cumbersome, uncomfortable and unhappy, and we can seriously injure ourselves if we fall over. If the Earth's gravity was only sixty percent of what it is now, we wouldn't have any of these issues. We'd not only fall over less frequently, but we'd do it without causing ourselves significant injury — just like most other creatures on Earth.

It would make perfect sense if Eden's gravity was forty percent less than it is on Earth. We would feel "at home" on Eden – light, comfortable and agile.

Mars is significantly smaller than the Earth, and it's almost certainly smaller than Eden too. Mars's gravity is sixty-two percent lower than the Earth's. So, while we would feel really light, we wouldn't necessarily feel comfortable. We might feel *too* light. We certainly wouldn't feel "at home" immediately; it would take us time to adapt, and the lower gravity might affect our health adversely.

Several studies have considered how living on Mars might affect our bodies. One major problem is that our bone density and muscle mass would reduce. This shouldn't affect us unduly if we remained on Mars permanently. But it would have a profound impact on our health if we stayed on Mars for more than a few months and then returned to Earth. Once we got back, we'd feel much heavier and more cumbersome than we do now, our muscles would struggle to support us, our bones would fracture easily, and we'd feel exhausted and in pain most of the time. Depending on how long we'd stayed on Mars, we might or might not recover from this. A long stay could lead to life-long health issues – to such an extent that returning to Mars might be the only feasible solution.

> Our bone density would start to reduce as soon as we arrived on Mars. In fact, it would start to reduce as soon as our spacecraft left the Earth on the way to Mars. The higher levels of calcium and other minerals on Mars would *eventually* make our bones stronger, but it might take several generations for this change to occur.

Let's look at the geology of Earth and Mars, and see if we can work out what it might be like on Eden:

Planet	Diameter (miles)	Circumference (miles)	Gravity (% of Earth)
Earth	7917	24901	1
Mars	4212	13263	0.38
Eden (estimated)	5992	18846	0.6

The figures for Eden are educated guesses based on our physiologies. We obviously don't know Eden's true size, density, or gravity. (Or where it is.)

Traveling from Eden to Mars would have less of an impact on our bodies than traveling from Earth to Mars (or from Eden to Earth). But the long-term effects are uncertain.

As we've just seen, it shouldn't be too much of a problem for those remain on Mars permanently. (And our chances of returning to Eden are pretty much zero anyway.)

But it's no wonder we feel so heavy and clumsy on Earth. No wonder we hurt ourselves when we fall over. No wonder we've developed thick bones and heavy musculature to compensate for the higher level of gravity. And no wonder we suffer from back problems and other issues as a result.

> There are several other serious issues we would have to contend with if we migrated from Earth to Mars. We'll take a more detailed look at some of these later in the book. But one of the main issues is that our brains would expand (and not in a good way). As a result, the pressure inside our heads could increase to a dangerous level.
>
> This is because blood and other fluids tend to collect in our heads in low-gravity environments, rather than being distributed evenly around our bodies. Our bodies have evolved significantly since we arrived on Earth hundreds of thousands of years ago, and one of the biggest changes has been an increase in our blood pressure. It would have been significantly lower when we first arrived on Earth. We wouldn't have been able to cope with the Earth's higher level of gravity for long periods, and we might have been prone to fainting if we stood up too quickly or for very long. That would have made us easy targets for wild animals – particularly the big cats – and the native hominins. This could explain why the first groups of modern humans to be brought to Earth failed to survive for very long.

If we spend very long on Mars, we might need to have pressure-release devices fitted in our heads or spines to avoid brain damage – or, at the very least, crippling headaches. The devices would allow some of the fluid to be drained out, and this would probably have to be done regularly, as it would keep building up.

3. Why Mars Suits Us Better Than The Earth

Children who are born on Mars might not need to have this done, but they would almost certainly have enlarged heads. This would be caused by the same condition that causes hydrocephalus or macrocephaly in children on Earth. But rather than recovering from these conditions, as most children on Earth do, they would retain their enlarged heads into adulthood and throughout their lives[3-10].

Within a few generations, the entire population of Mars might have enlarged heads. But at least they would no longer need to have their excess fluid drained out regularly.

The big disadvantage of having larger heads is that giving birth would become even more difficult. It's bad enough when children have normal-sized heads. The lower gravity on Mars would compensate for this to some extent, of course, as everyone would have less musculature and women's birth canals should become larger and more easily able to expand.

But it's not all bad news. An advantage of having larger heads is that the white matter in our brains should have room to expand. This would allow different parts of the brain to communicate more easily. As a result, future generations of Martians should be able to solve complex problems more easily than people on Earth.

> The people of Mars wouldn't necessarily be able to think faster than the people on Earth, though. The pathways between their neurons would be longer, so it would take slightly longer for signals to pass along them.

> With their enlarged heads, the future human inhabitants of Mars might come to resemble the Grey aliens we're so familiar with. We can also assume that the Greys will be able to solve complex problems more easily than us. But we should be able to think a little faster than them.

• Our 25-hour body clock

You'll have noticed that the days on Earth feel too short, and there are never enough hours in the day. There's a good reason for this. In numerous experiments, sleep researchers have found that our internal body clocks measure days in periods of around twenty-five hours, not twenty-four hours.

Our body clocks reset themselves each morning when they're exposed to daylight, but this can cause a severe shock to our systems. It's no surprise that this is the time of day when most heart attacks occur.

Even if we survive waking up, we feel tired all day, we rush to get everything done in the limited time available, and we struggle to get by on too little sleep. Once again, this is not a consequence of modern living: it's been that way ever since our ancestors first arrived on Earth hundreds of thousands of years ago. And, of course, it's yet another factor that contributes to our chronic ill health. An extra hour each day would make all the difference in the world.

Clearly, a day on Eden must last around twenty-five Earth hours.

But Mars isn't far behind. A day there lasts for 24 hours and 39 minutes, so we could live there perfectly happily. The extra 43 minutes a day (compared with the Earth's 23 hours and 56 minutes) would be extremely welcome.

> Interestingly, it may be possible to change the way our internal body clocks measure the length of a day. Geneticists have managed to do it in fruit flies by manipulating their genes. But they haven't conducted any human trials yet. It's a shame the aliens didn't do this to us when they brought us to Earth and we lost over an hour each day.
>
> If the aliens had taken us to Mars instead of bringing us to Earth, there should have been no need to modify our body clocks. There's only a twenty-one-minute difference between the length of a day on Eden and the length of a day on Mars.

- **Our 23-month body clock**

We age in Mars years (22.6 Earth months) rather than Earth years (12 months) – in our minds at least[3-11]. According to *Medical Xpress*, for every decade that passes, we feel we're only five or six years older.

The University of Virginia in the USA carried out a study in 2018 that confirmed this. Their researchers found that once we pass the age of twenty-five, we think of ourselves as being younger than we actually are.

This is further confirmation that Eden's speed of rotation and orbit must be similar to Mars's.

> Interestingly, the younger we *feel* we are, the less likely we are to experience health issues such as cognitive impairment, dementia, depression, diabetes, and hypertension.

> Although the studies confirm we feel younger than we actually are *on average*, it doesn't affect everyone the same way, and there are some anomalies. People with chronic health issues may feel *older* than they actually are. I've also come across a few people who say they feel *really old*, even though they're actually quite young. We may need to carry out further studies on this. They might have hidden health issues they're not aware of. Or it might be something to do with their mental attitude or upbringing.

- **We survive, but we don't thrive**

Most of us survive rather than thrive, and we live our lives day by day, just about getting by. Sadly, many of us don't enjoy our lives. Again, this is not a consequence of modern living; it's been that way ever since we arrived on Earth. Remember that we lived in survival mode for hundreds of thousands of years. This might well be a hangover from that time.

It's not that we hate our jobs, if we have them. We could easily get different jobs if we wanted to. And it's not about having to work for a living, or having no time to enjoy our lives or hobbies. Wealthy and retired people with no need to work often feel exactly the same as those who have jobs.

We could easily move to a different country if we wanted to. But that still wouldn't change the way we feel overall.

The problem is that we don't fit in on this planet; we're out of place. The days are too short and their length keeps changing, meaning that our body clocks are permanently screwed up. The gravity is too strong so we feel heavy and clumsy, and we ache all over. The sunlight is too strong and it makes us sick. The food and water are wrong; they aren't what we're used to, and the food doesn't contain enough minerals. The overall environment doesn't suit us – regardless of where we live on the planet. The summers are too hot, the winters are too cold, and we either want to migrate or hibernate. I could go on.

There's no question that we would feel significantly better if we lived on Mars. If it was habitable, and if crops from Eden grew there, every one of these issues would disappear. And we *would* undoubtedly thrive and enjoy our lives.

- **We can't sense natural disasters**

There are all sorts of natural disasters on Earth. Most animals can detect them, and they leave the area, or take precautionary measures, long before they occur. But we have no inherent way of detecting them, and we often get no advance warning.

We can't sense impending earthquakes, tsunamis, volcanic eruptions, hurricanes, cyclones, monsoons, floods, wildfires, or any other kind of natural disaster.

In recent times, we've developed scientific instruments that can give us an early warning of *some* of these disasters. But the Earth's native creatures are still well ahead of us. Sometimes, natural disasters happen with no warning whatsoever – or so it seems. Yet when they occur, the other animals are nowhere to be seen. They sensed the coming disaster and made their escape hours – or sometimes days – beforehand[3-12]. Yet our most sensitive scientific instruments and data analysis systems detected nothing amiss until the moment disaster struck.

The reason why we can't sense these disasters is probably because they don't occur on Eden. We never evolved the ability to sense them because there was no need for us to do so.

There aren't many natural disasters on Mars either. Minor marsquakes[3-13] occur from time to time, but they're relatively harmless. They're probably caused by frozen water shifting around in underground fissures[3-14], or by small meteorites striking the surface[3-15].

As there are no tectonic plates on Mars, there are no major earthquakes or tsunamis. There are several volcanoes, but they became extinct long ago. Mars's crust and mantel have cooled, solidified and become too thick for the molten rock deep inside the planet to reach the surface[3-16].

Mars does, however, suffer from global dust storms. These could (and already do) cause us problems by disrupting our communications and blocking our solar power arrays. The particles could get into our machinery, electronics, and clothing, and cause us all sorts of issues. But this is only a problem on present-day Mars, which has no surface water. Dust storms wouldn't have been a problem in the past when the surface and atmosphere were moist, and they shouldn't be a problem if we terraform Mars in the future.

> Some researchers believe the global dust storms may have carried much of Mars's surface water into the upper atmosphere. Once it was there, it could have been more easily broken down by particles from the Sun, and carried away into space[3-17]. Although there's still plenty of water underground on Mars today, a body of water the size of an ocean is missing. This could explain what happened to it.

A terraformed Mars *might* suffer from other weather-related issues, such as storms and floods, but we just don't know at this stage. If they occurred, we might be able to adjust the climate to prevent them or lessen their impact. We'll come back to this again in Chapter 9, when we take a look at how we might terraform Mars.

Natural wildfires would be unlikely, as the amount of solar radiation reaching Mars is fairly low. But, depending on the oxygen level, humans could start accidental (or deliberate) wildfires.

> The technology we develop to terraform Mars could also be used to alter the Earth's climate. We should be able to find ways of reversing or limiting the impact of climate change. We should also be able to prevent weather-related disasters such as hurricanes, cyclones and storms, floods, droughts, and so on. But we probably won't be able to prevent geological disasters such as earthquakes, volcanic eruptions, or tsunamis.

- **Our poor sense of direction**

Many of the Earth's native creatures find their way around by sensing the planet's magnetic field. Migratory birds, and some fish and amphibians[3-18] can do this. But we can't.

You might think we didn't develop the ability to sense magnetic fields because Eden didn't have one. But, as we've already seen, it's highly unlikely that life could survive on a planet that didn't have a magnetic field. Not only would it lose its atmosphere and water, the surface would be sterilized by radiation and energy bursts from solar flares.

> Astronauts on Mars who were unable to shelter from a solar flare might not die from radiation burns and sickness straight away, but they would undoubtedly suffer from long-term health issues. Cancer and cardiovascular disease would reduce their lifespans. Hopefully, in the event of a large solar flare being detected, they would receive sufficient warning and be able to take shelter deep underground.

Researchers have found a small group of cells in our brains that can detect magnetic fields, but we can't access any of the data from them. The cells could be a left-over remnant or vestigial feature from when we evolved from an earlier species. But why would *that* species have been able to sense the magnetic field if they evolved on the same planet as us? And why have we lost the ability to sense it when it would have been so useful to us?

I can think of two possible scenarios:

Eden had a strong magnetic field that the group of cells in our brains could detect. When we were brought to Earth, which has a weaker magnetic field, the cells could barely detect it. We developed other ways of finding our way around, and our ability to sense magnetic fields was lost.

Alternatively, the aliens that brought us to Earth may have blocked our ability to sense its magnetic field. For example, they might have severed the connection between the magnetism-sensing cells and the part of our brain that processed the data. Perhaps they didn't want us to move around too much. Fortunately, we're a more resourceful species than they may have realized.

There's a remote possibility that Eden doesn't have a magnetic field. For example, it might orbit a small or dim star – such as a red dwarf – that might not emit much solar wind or radiation. We should be able to survive there even if Eden doesn't have a magnetosphere to protect us. But why, in that case, do we have cells in our brains that can detect magnetic fields? My best guess is that we inherited them from the native hominins on Earth when their genes were spliced into us or when we interbred with them.

> Does that mean the Earth's native hominins could find their way around using Earth's magnetic field? I'd love to know if they could, but it's something we'll probably never find out.

> Ever since the researchers announced that they'd discovered the magnetism-sensing cells in our brains, I've wondered whether we might find a way of reactivating them. I don't believe we will, because our technology has replaced and superseded our need for them. We've had sextants and compasses for hundreds of years, and we now have satellite navigation systems in our vehicles and phones. Companies such as Google have developed wearable guidance systems – smart glasses – that can tell us where we are and which way we need to go. Future versions might be implanted directly into our bodies as part of a multifunction upgrade package.

> Looking even further into the future, there might be no need to implant these systems. Wireless sensors on every street might read our thoughts and transmit data directly into our brains in a form we can understand. In effect, we would all become artificially telepathic.

Mars currently has no magnetic field, and it hasn't had one for at least four billion years. As it's smaller than the Earth, its magnetic field was probably no greater than the one that surrounds us today, so it's unlikely that the cells in our brains could have sensed it.

The first step in terraforming Mars should be to give it an artificial magnetic field. This is essential, because otherwise the newly restored atmosphere will be eroded by the solar wind, and the surface will be exposed to high levels of radiation. An artificial magnetic field (or magnetosphere) will divert the solar wind and radiation around the planet.

It's unlikely that the cells in our brains will be able to detect this magnetic field, or use it for navigation. But we won't need to anyway, because we'll use the same guidance systems that we have on Earth.

> There are other ways of finding your way around. Some migratory birds navigate using polarized light – as well as by sensing the Earth's magnetic field. Some species of fish can find their way back to their spawning grounds using their sense of smell. We can't do either of those things. But, interestingly, some species of beetle navigate using the stars – and that *is* something we can do.

- **Lack of defense mechanisms and predators**

We lack sharp teeth and claws, so we would have been unable to defend ourselves against the big cats that live on Earth – particularly in east Africa where we supposedly evolved. When we first arrived, there would have been many more big cats than there are today. We couldn't outrun them, as most of them are faster than us. Nor could we climb trees to escape them, as many of them can do that too. All we could do was to wave branches or burning sticks, or throw rocks at them, in the hope of driving them away. Fortunately, that seems to have worked most of the time.

> The timeline of human evolution on Earth – which we'll look at in Chapter 10 – shows that we became an endangered species on more than one occasion. Most anthropologists believe this was because of a series of natural disasters. But it's possible that huge numbers of people were killed by wild animals.

Our lack of defensive mechanisms suggests we evolved on a planet where we had no predators. This would be the case on Mars as well as on Eden. In fact, if we had been taken to Mars, we might have been the only mammals on the planet.

> This would have presented our biologists with a massive conundrum. The Theory of Evolution wouldn't have existed, as there would have been nothing we could have evolved from. The only possible explanation for our presence on Mars would have been that we had been brought there

> from somewhere else, or that we had been created and placed there by a deity.
>
> But I'm sure our scientists would have come up with plenty of alternative theories.

• Overpopulation

A species becomes overpopulated when there's an abundance (or overabundance) of food, and a lack of disease, predators, and other life-limiting factors.

There's plenty of food on Earth (though it's not fairly distributed), most countries have good health care programs, and we have plenty of weapons to defend ourselves against the few remaining predators (and each other). So our population has exploded.

> It's likely that the food supply on Eden was more limited in availability, and it was probably also less nutritious. This would have limited our lifespans, kept our sperm counts low, and prevented overpopulation. On the plus side though, as Eden doesn't have seasons, what little food there was would probably have been available all year round.
>
> One of the biggest problems with Mars is that the winters last for nearly six months. That's a long time to go without food, and it would have been difficult to store enough to last us that length of time.

> We won't have that problem when we establish communities on Mars in the future. We'll be able to grow food all year round in hydroponic farms underground.

If our harvests on Eden had failed for any reason, only a few thousand of us might have survived. That's barely enough to sustain a viable population. It's unlikely that our population size was ever very large in any case, as the lack of available food would have kept us in check. We may have been on the verge of extinction when we were brought to Earth.

Once we've terraformed Mars, it will still have seasons that last almost six months. One way around this would be to work with the seasons, as birds and some nomadic tribes do on Earth. We might plant crops in the north during the spring months, then venture south to harvest the crops we'd planted there the previous fall, then migrate back to the north to harvest the crops we'd planted there in the spring, and so on.

We may have been brought to Earth because Eden's climate was becoming warmer. That might have been because its star was becoming more luminous – just as the Sun will in the second half of its life. But we shouldn't be too surprised if it turns out to be because of something *we* did.

One consequence of the warmer climate is that more food may have become available. This may have led to a population boom. But the warmer climate may have caused droughts and famines, and there may have been *less* food available. This could have caused a substantial *decline* in our population, and put us on the verge of extinction.

Eden's cool climate might also have kept pathogens such as viruses in check. As the climate warmed, infections may have become more frequent and more severe. Since there are no seasons on Eden, the pathogens would have been able to infect us all year round, and their numbers would never decrease. Again, we might have been on the verge of extinction when the aliens rescued us.

- **Our reaction to carbon dioxide**

No one has been able to explain why a rise in the carbon dioxide level causes us to go to sleep – or sends us into a deeper sleep – yet it causes the other animals on Earth to wake up. The carbon dioxide level rises during things like fires and volcanic eruptions. Other animals sense it, wake up, and escape to safety – as long as they aren't trapped in our homes. Before the advent of smoke detectors, millions of people died when their homes caught on fire. But many people were saved because their pet dogs woke them up.

Why we react in the opposite way to every other species on Earth is an interesting question – and a strange and dangerous anomaly. I can only assume the carbon dioxide level on Eden must be so low that we can barely sense it. Or perhaps there are no fires there (because the oxygen level is much lower) and no volcanic eruptions, so the carbon dioxide level never changes.

> When we terraform Mars in the future, it might not have as much carbon dioxide in its atmosphere as the Earth's does. We'll look at this in more detail in Chapter 9.

- **Children versus plants**

Newly born animals can identify which plants are safe to eat and which ones are poisonous or taste horrible. They have an innate ability to do this, and they don't have to be taught it. But our children lack this ability. They have food neophobia – a fear of new or unknown foods[3-19].

Many of them refuse to eat vegetables, saying they taste "poisonous" or "yucky." Yet they'll happily put things like poison ivy in their mouths if we don't keep a close eye on them.

This is bizarre and dangerous behavior. It suggests that our children can't recognize the plants on Earth, whereas every other animal can. Our children have to be taught this information – and even then, many of them reject it.

We can assume that our children *would* have recognized the plants on Eden, and they would have known instinctively what they could and couldn't eat. This knowledge might have been passed down to them from their ancestors in their genetic memories, just as happens in other animals.

Some of our ancestors may have arrived on Earth over 400,000 years ago, but that clearly isn't long enough for our genetic memories to have adapted. We *still* can't recognize which plants are good to eat unless someone teaches us. I wonder how long the process takes? Millions of years perhaps?

Would we have recognized the plants on Mars if we'd been taken there instead of being brought to Earth? If those plants were native to Mars and not found on Eden, then no, we would not have done. We would have faced the same issue that we do on Earth today. The only way we would be able to recognize them is if the aliens had transplanted them from Eden.

OBJECTION: This is simply down to the wide variety of food we eat, and the fact that we've modified it so much that it no longer resembles the natural species that evolved here.

No, this isn't true. We can see the same behavior in children all over the world, regardless of whether they live in developed or developing countries, and regardless of whether they eat modified or natural species, and a wide or narrow range of foods.

Another strange anomaly is that most children prefer sugary foods, whereas most adults prefer foods that are less sugary and more bitter. Our tastes change as we get older. We don't see this in any other species; once they've been weaned, young animals generally eat the same food as their parents throughout their entire lives. This is difficult to explain. It would make sense if we were *all* attracted to sugary, energy-rich food, because it would have been scarce on Eden. But why would we develop a preference for less-energy-rich food as we get older?[3-20]

Some people *never* lose their sweet tooth, of course. They crave energy-rich foods throughout their lives. As it's widely available in most countries, they can become morbidly obese as a result, suffer from serious health problems, and die younger than those who prefer less sugary foods.

> This undoubtedly comes back to the fact that energy-rich food was scarce on Eden. But why have some people lost this craving while others have retained it? It could be a genetic issue, or it could be a behavioral one.
>
> Interestingly, we see exactly the same thing in people in developing countries where food is scarce. One would expect *everyone* in those countries to have a sweet tooth throughout their lives, but they don't.

When we terraform Mars, space for growing food will be at a premium for the first few generations. It will make sense to grow a small range of food that's energy- and nutrient-rich. That means it will probably taste sweet, so the early settlers might need a sweet tooth throughout their lives too.

Future generations will have more space for growing food, and they'll be able to grow a wider variety of crops, including varieties that are less sugary.

• Birth issues

Our babies' heads are almost too large to squeeze through their mothers' pelvic openings. There's no room to spare, and labor can last for twenty-four hours or more – far longer than in any other animal. No other non-domesticated animal has such problems.

At the same time, our babies remain helpless long after they're born – much more so than other animals on Earth. They also have longer childhoods than any other species.

Our babies couldn't remain inside their mothers for any longer because their heads would become too large to pass through their pelvic openings, and their energy demands would outstrip their mothers' ability to supply it.

The main issue here is that the Earth's gravity is higher than we evolved to cope with. We evolved on Eden over billions of years. In the 400,000 years since our earliest ancestors arrived on Earth, we've had to develop thick bones and heavy musculature to support ourselves. That's a *really* short timescale – it's more of a "hack" than an evolutionary development. We'll probably adapt to living here eventually, but it might take us millions of years. And if nature doesn't deliver the goods, technology undoubtedly will.

We saw earlier that our heads become larger in lower gravity. It's highly likely that our heads were larger than they are now when we first arrived here from Eden. But we would also have had thinner bones and less musculature, so we would have been more flexible than we are now.

> As we started to adapt to the higher level of gravity on Earth, and we developed thicker bones and heavier musculature, more mothers and babies would have died during childbirth. Their babies would still have had large heads because they evolved on a planet with lower gravity, but the pelvic openings they had to pass through would have become smaller, tighter and less able to expand.

Over time, the size of our babies' heads has reduced. Nature favored those with smaller heads, and more of them survived. Also, unlike other species on Earth, our babies' craniums can compact during childbirth, as they have bony plates that can overlap. The plates slip back into place once the baby is born, and they fuse into a single solid piece during early childhood.

This process could not have evolved in the 400,000 years since we arrived on Earth – it would have taken millions of years. So it must have been something that also happened when we lived on Eden.

> The latest research shows that our babies' heads are *still* becoming smaller. So we obviously haven't finished adapting to life on Earth yet.

We would also have large heads if we lived on Mars. In fact, they might become even larger on Mars than they were on Eden, because Mars's gravity is lower. You might think this would make giving birth more difficult – and it absolutely would for the first few generations. But it would eventually become easier because our bodies would return to their natural form, with thinner bones and lighter musculature. Women would have larger and more flexible pelvic openings, and their babies would be able to pass through them with ease.

> As human babies have overlapping bony plates in their craniums, it's possible that those born on Mars might have heads that are no larger than those born on Earth. They will almost certainly *end up* with larger heads than people born on Earth, because the lower gravity will encourage fluid to build up there after they're born and as

> they develop into children and adults. But childbirth should be no more difficult than it is here on Earth, and it might even be significantly easier.

These changes won't happen immediately; it might take hundreds of thousands of years for the adaptations that have occurred on Earth to be reversed.

It will take even longer for the migrants from Earth to adapt fully to the lower gravity on Mars.

But thinner bones, lighter musculature, and larger heads aren't the only changes we can expect to see. There will be many other changes too – as we'll see later in the book.

Another reason why we might have large heads is because we were hybridized with the aliens themselves. As we've already noted, most aliens have large heads too, suggesting that they come from planets with low levels of gravity. If they spliced part of their genome into ours, this could explain why our heads have become so large that they can barely fit through our mothers' pelvises.

Had we been taken straight to Mars rather than being brought to Earth, we wouldn't have had these problems. The gravity on Mars is lower than it is on Eden, so our bones would have been smaller, and our musculature would have been lighter and more flexible. We might have been smaller overall too, because of the lower amounts of nutrition available (assuming we ate the same foods as we did on Eden).

- **Morning sickness**

No one knows what causes morning sickness in pregnant women. As far as we're aware, no other animals suffer from it.

The most likely reason is that the Earth's food, water and environment don't suit us and we need to flush out the toxins to make our bodies as safe as possible for our babies.

Pregnancy is often associated with strange food cravings. These are almost certainly our bodies' desperate attempts to obtain the vitamins and minerals they lack – and which they can't get from our regular diet.

There's also the issue of genetic incompatibility between mothers and their babies, especially if the baby has rhesus positive blood and its mother doesn't. Again, no one knows why some of us are rhesus positive and some are rhesus negative, or why the two types aren't completely biologically compatible. You can't tell which type someone is without analyzing their blood, which makes things even more difficult. It would certainly help if the two types looked (or smelled) different.

> If a woman with rhesus negative blood becomes pregnant with a rhesus positive child, and if any D antigens from that child's blood get into her bloodstream, her body will reject any subsequent children that also have rhesus positive blood. Antibodies in her blood will attack the fetus's red blood cells, causing hemolytic disease, which leads to anemia and jaundice.

> Women with rhesus negative blood are routinely given an "Anti-D" injection during their first pregnancy if their partner has rhesus positive blood. This mops up any D antigens that cross over from the fetus, and prevents her body from becoming sensitized to them.

By itself, this doesn't prove we came from another planet, or that external forces adapted us to live on Earth, Mars, or anywhere else. But it *does* suggest that humans might be two distinct sub-species, or that we're a hybrid race.

It also proves we can't all have evolved from a single African ancestor, as mainstream anthropologists insist. If we had, then we would all be fully compatible with each other. But we're definitely not.

Another possible cause of morning sickness is the significant amount of genetic interference the aliens have caused in our bodies. This might also be a significant factor in many of the other chronic illnesses and conditions we suffer from.

Pregnant women are unlikely to have suffered from morning sickness or unusual cravings on Eden. They evolved there, so they shouldn't have had any toxins in their bodies to get rid of, and they should have been able to get all the vitamins and minerals they needed.

If Mars's mineral composition is similar to Eden's then it shouldn't be a problem there either.

Supporting evidence

The evidence in this next section provides further confirmation that we didn't evolve on Earth, and corroborates the evidence from the first section. But it doesn't necessarily pinpoint the Earth, Mars, or anywhere else as our intended home.

This is more-or-less the same evidence that I presented in *Humans Are Not From Earth*. But as it doesn't relate only to Mars, I've summarized most of it. I've also added some new ideas and theories that have been put forward since that book was published.

- **Concealed ovulation**

Fully concealed ovulation is rare in primates, and even more so in other mammals. So it's strange that ours is concealed so completely that even the woman herself is often unaware it's happening.

This strongly suggests we may have evolved somewhere else. Perhaps concealed ovulation is the norm on Eden.

We have no idea what the norm would have been on Mars if it had developed advanced life. But if we had been the only mammals there, *our* norm would have been the Martian norm.

- **Violence**

Our innate violence could be the reason – or one of the main reasons – why we were removed or expelled from Eden.

The big question is whether we were deliberately created to be violent, or whether we became violent all by ourselves.

Later in the book, we'll look at the theory that we were created that way. But it's possible that we evolved that way because it made us better able to compete for food, territory, breeding partners, and so on. Our most violent ancestors would have gotten the most food, the biggest and best territory, and the best breeding partners. So violence would have become an inherent part of our natures. We can see the same behavior in other species – but we seem to have taken it to a whole new level.

Would we have been any less violent if we'd been taken to Mars or any other planet, rather than the Earth? The short answer is, of course, no.

> If some of us had been taken to Mars and some to Earth, the two groups would almost certainly have declared war on each other. In fact, this might well happen in the future when settlers from Earth migrate to Mars and colonize it[3-21]. Once they become fully self-sufficient, the settlers will want full independence too, including freedom from Earth's control, and the right to establish their own government, laws, taxation, trade tariffs, and so on. Earth, on the other hand, will probably want to retain ownership and control of Mars. So we might see a space-based equivalent of the American Revolutionary War.

> The American Revolutionary War began 155 years after the first English settlers arrived in what would become the USA. I wouldn't be too surprised if the timescale on Mars was similar. The Martian Revolutionary War might begin

> around 155 years after the first permanent settlers arrive from Earth. I have no idea who would win. But I'm sure the Martians *will* eventually gain their independence – even if they lose the war the first time around.

> There were 102 settlers on the *Mayflower* ship that colonized America. Researchers believe we could colonize Mars with a similar number of people – about 110[3-22][3-23].

- **Selfishness and greed**

We are the only species on Earth that's *motivated* by selfishness and greed. We aren't the only species that's selfish or greedy, but we are the only species that's *motivated* by these things. This leads to things like distrust, envy, hatred, spite, and (of course) violence.

Our selfishness and greed are so ingrained that they must have existed long before we came to Earth. Our motivation to behave this way must have evolved (or been created) on Eden.

Would we have been motivated by these things if we'd been taken to Mars instead? Again, the answer is yes.

Our motivation to behave this way is inherently linked to our violence, as we saw above. The most violent of us got the best food, territory, breeding partners, and so on, and this evolved into an innate (genetic) motivation to always want *more* – at any price. As a result, *all of us** are permanently ready, willing and able to fight for the things we want.

*Apart from the most seriously infirm and disabled, of course — and even *they* can put up a heck of a fight to get what they want.

• Destruction of our resources and our environment

We are the only species on Earth that deliberately destroys the environment through our natural behavior. We're also the only species that recognizes and understands we're destroying the environment, yet continues to do so.

There's a strong connection between this factor and the previous one. The worst damage is caused by powerful and wealthy individuals and their global corporations because they and their shareholders always want *more* (of everything, but mostly money and power). They lobby and bribe politicians and political parties to get what they want, regardless of the damage their operations do to the planet.

But the same thing happens on a smaller scale too. We can see examples of vandalism, graffiti, littering, and destruction in almost every community in the world, whether large or small.

We would, of course, have done these terrible things on Eden too. It might be another reason why we were expelled from it. And we would have done exactly the same thing if we'd been taken to Mars or any other planet.

> One thing is certain: future settlers on Mars will vandalize their new planet, no matter how well they're vetted before setting off from Earth, and no matter how well their behavior is policed.

It's worth noting that things have started to improve since around 2016. Environmental campaigners have protested about the damage and destruction for decades, but they now have widespread public support and powerful backers. Governments are finally listening to them, and they're being forced to take positive action.

Much of that action is being led by the younger generation, who are taught about the issues in school and fear for their futures. Previous generations were warned there would be serious global problems *in the future* if they didn't change their behavior. But as far as they were concerned, that future was a long way off and the problems were too big for ordinary people to get involved in and make any difference. So they *didn't* change their behavior. But those problems are no longer *future* problems, they're *today's* problems.

- **Super-intelligence**

Hominins' brains increased in size significantly, starting around 800,000 years ago and ending around 200,000 years ago. This is strange, because that degree of evolution should have taken millions of years.

When the hominins first appeared on Earth around 5.8 million years ago, their brains were the same size as a chimpanzee's, and they remained that size for five million years. We have no idea what triggered the massive growth in brain size and intelligence, or why it was compressed into just 600,000 years. Some researchers have suggested it may have been due to climate change. But it's more likely the work of an external influence. The aliens who brought us here may have carried out their first hybridization experiments on the Earth's native hominins before they tried hybridizing us.

> It's also possible that the aliens made the native hominins stronger, more agile, and more intelligent so they had a better chance of surviving once we arrived. It didn't work, of course.

All of this suggests that the aliens were interfering with life on multiple planets: definitely the Earth, Mars and Eden, but almost certainly other planets too.

The fact that we are both violent *and* intelligent makes us doubly dangerous, of course. That's even more reason to forcibly migrate us to somewhere far away from the other civilizations.

- **DNA anomalies**

Geneticists working on the Human Genome Project found 223 "orphan" genes (out of our surprisingly small total of around 20,000 genes) that they can't explain. These genes aren't found in any other species on Earth. Some researchers believe the aliens may have spliced them into our genomes to help us survive on Earth after their early attempts to rehouse us here failed. But it's more likely that we brought them with us from Eden. Several of them are undoubtedly connected with our violence.

If we didn't bring them with us, then we have no idea where they came from. They might have been created artificially or taken from another species – perhaps even from the aliens themselves. Our own geneticists can already create artificial genes, and the aliens' technology will be millions of years ahead of ours, so there's no reason why they couldn't be synthetic.

As we saw earlier, our geneticists have also found scars in our genome where sections of DNA have been added or removed. They've also found numerous other anomalies. This provides us with clear evidence that our genome has been manipulated by external forces.

> Lloyd Pye covered this in detail in his book *Everything You Know Is Wrong: Human Origins*.

- **Semi-aquatic features**

Humans are the only land mammals on Earth that have subcutaneous fat. This is normally only found in marine mammals. We also have numerous other features that suggest we went through an aquatic or semi-aquatic phase during our evolution.

If we look at the fossil record for Earth's native hominins (from whom we supposedly evolved), there's no evidence at all of an aquatic or semi-aquatic phase. Nor is there any trace of it in the fossil record for the primates and lower mammals that came before them. So where did these features come from?

My hypothesis is that we *did* have a semi-aquatic phase during our evolution, but it occurred on Eden, not on Earth.

The "aquatic ape hypothesis" has been officially disproved on Earth because these features aren't present in our supposed ancestors. But there's no other way to explain our streamlined body shape, our descended larynx, our diving reflex, our need for DHA (an omega 3 fatty acid that's abundant in seafood but rare on land), our affinity for water, our mating position, our ability to hear better underwater than we can on land, the vestigial webbing that sometimes occurs between our fingers and toes, our kidney structure, and the vernix (waxy coating) that covers our babies when they're born. These features occur in several marine mammals, but they're unknown in other land mammals.

Some researchers believe our aquatic evolutionary phase also goes some way toward explaining our hairlessness, the complexity of our brains, and our bipedalism.

You'll find more information about this in the first book in the series, *Humans Are Not From Earth*, along with details of why we couldn't have evolved from the marine mammals on Earth.

> Some researchers claim that fossil evidence of our aquatic past *has* been discovered in coastal areas on Earth. But they say mainstream scientists ignore it because it doesn't fit with their conventional theory of evolution.

• The shape of our skulls

Our skulls look nothing like those of our supposed ancestors, including *Homo erectus* and *Homo heidelbergensis*. They also look nothing like our supposed closest hominin relatives, the Neanderthals and Denisovans. Our skulls are different in just about every aspect – and we also have chins, which none of the other species have.

The hominins' skulls, on the other hand, look much like each other and the hominins that preceded them, and the evolutionary line is clear and obvious. There's no doubt that these species are related, evolved from one other, and belong on Earth.

If we look at our own skulls, and the fossil record for the other bones in our bodies, it's clear that we have no place in the Earth's evolutionary timeline. There's certainly no evidence to suggest that we evolved from any of the native hominins.

• The lack of a penis bone (baculum)

Most mammals and all of the great apes, including the chimpanzees and gorillas, have penis bones. But we don't. This is a good thing because it means we can achieve around 250 sexual positions, whereas other land mammals have to settle for just one. But no one has been able to explain why we don't have them.

> There are plenty of theories though.

• Gender identification issues: non-binary

Recent social changes have brought to light a considerable number of people who are biologically neither male nor female, or who are both male and female, or they identify as such.

Some commentators have suggested that this is a random developmental defect. Some animals mimic the behavior of other genders, and the clown anemone fish can actually switch genders, so this doesn't necessarily indicate we're from another planet. We also have no idea whether it's a recent phenomenon or whether it's as old as (or older than) humanity itself.

It could be a further sign that our genomes have been manipulated, and that perhaps things didn't go entirely to plan. Perhaps the aliens who meddled with our DNA didn't understand what they were doing quite as well as they thought they did.

> Since they're undoubtedly millions of years ahead of us, the aliens must be completely familiar with *their own* genomes, and they can probably manipulate it in any way they wish. They might also understand and be able to manipulate the genomes of every other living thing on their own planet. But it's unlikely that we are from their world, or anywhere close to it. So the techniques they've developed might not work so well on us.

There's another theory that our DNA might be deteriorating, perhaps because of chemical and radioactive contamination in our environment. This could explain why the number of people affected by this issue (as well as the many chronic illnesses we looked at earlier) appears to be increasing.

But we should also consider the possibility that our DNA's deterioration might be planned and deliberate. It might also be connected to the fact that the male sperm count is falling. This could indicate that Mother Nature is trying to reduce the size of our population to a more sustainable level. But it might be the case that an alien civilization is doing it in order to save the planet.

The problem is if it continues unchecked, there's a real possibility that it could lead to our extinction in a few hundred years.

- **Our *real* missing link**

If we had evolved on Earth, our immediate ancestor would probably be the same as the Neanderthals' and Denisovans' ancestor: *Homo heidelbergensis*. But as we've already seen, if we compare the skull of *H. heidelbergensis* with our own, it's obvious at a glance that we could not have descended from them. They're significantly different in shape, size, structure, and weight. *H. heidelbergensis* skulls are much heavier and more robust than ours, with pronounced brows and inward-facing lower jaws. The structure of their sinuses is significantly different too.

If we *had* evolved from *Homo heidelbergensis*, it would have taken several intermediate stages for us to get to where we are now. We haven't found any evidence of any intermediate species, either as physical remains or as remnants within our DNA. And, in any case, there isn't enough room in the timeline of human evolution for any intermediate species.

Another issue, that's rarely mentioned, is the fact that all of these species, including modern humans, were present on Earth *at the same time* around 400,000 years ago. The likelihood that we evolved from any of them is therefore zero.

We must have evolved on another planet, or we must have been created as a hybrid species.

If we're hybrids, then we might be a mixture of the aliens and one or more hominins from Earth. But it's more likely that we're a mixture of our original selves (as we were when we lived on Eden) and one or more hominins from Earth.

- **Feeling out of place**
- **Depression and unhappiness**
- **Self-destruction**

Many of us have a deep-rooted feeling that we don't belong on Earth. It's not just that we feel we're living in the wrong city or the wrong country; we'd feel exactly the same wherever we lived on Earth.

> Many indigenous tribes, including some that had no previous contact with the outside world, point to specific stars and say: "Our ancestors came from there." It's part of their identity and culture.

Some people describe this deep-rooted feeling as the type of sadness you feel when you're missing family members you haven't seen in a while. Others say it's like a constant but distant sorrow. A longing for ... something ... that can never be satisfied. A relationship you yearn for, but you know in your heart can never be fulfilled. Or the feeling of melancholy that comes from knowing you will never see your true home again.

I believe this goes a considerable way toward explaining why so many of us are unhappy or depressed. It's not just a factor of modern living, as some readers have suggested. For example, many people say they'll only be truly happy when they retire from their jobs. But when they retire, they realize they now have nothing to live for. Their jobs were the only things that were keeping them going – even if they hated them. They believe they can no longer make an active and meaningful contribution to society. So they fall into a deep depression and die early.

> They could *easily* make an active and meaningful contribution to society by helping a community cause or charity. Many of these organizations depend on volunteers to survive, and they need all the help they can get. The volunteers not only play a vital role in keeping the organizations running, they also build a support network of friends and knowledgeable people they can call upon if they need help themselves. So they *can* play a useful, active and meaningful role in society, and it's a great cure for loneliness too.

As far as I can tell, there has *never* been a time in our history when the majority of us were happy. Unfortunately, our unhappiness and depression, together with our inherent violence, mean we are tremendously self-destructive. According to the psychologist Sigmund Freud, we have: "an innate death drive that impels us to pursue our own downfall." No other species on Earth has this.

Would we feel more at home on Mars? It's hard to know. If it closely resembled Eden in *every* significant way, then it's certainly possible.

Our descendants will find out what it's like to live on Mars. But first, we'll need to fully terraform it, re-establish its atmosphere and surface water, and make it self-sustaining. It might take 100,000 years or more to do this, but I'm sure we'll accomplish it eventually. In fact, as we'll see later in the book, our survival as a species *depends* on it.

In the meantime, we'll continue adapting and evolving on Earth, and we'll become less and less like the species we were when we first arrived.

It's important to remember that the first groups of modern humans who were brought to Earth died out very quickly. The aliens adapted our genomes (or borrowed genes from the native hominins) to make us more suited to living here. But those adaptations would have made us *less* suited to living on Eden and less suited to living on Mars.

It might take us 100,000 years or so to terraform Mars, but we'll continue evolving here on Earth during that time. So, while we'll become better adapted to living on *this* planet, we'll find it harder to live on Mars, even if we can turn it into an exact duplicate of our original home planet, Eden.

> Mars can never be an *exact* duplicate of Eden, because Mars is almost certainly smaller and its gravity will be lower.

We will eventually adapt to living on Mars, of course, just as we'll eventually adapt to living on Earth if we remain here. But it might take us hundreds of thousands of years, or even millions of years. And our bodies will change so radically that we'll effectively become a different species.

Once we've gone through that process, we'll find it all but impossible to live on Earth if we ever wanted to return. Even a short visit would be painful and potentially deadly. The higher gravity and air pressure would be unbearable, and it would put a tremendous amount of stress on our hearts and bodies.

> The citizens of Mars would also look different from everyone who remained on Earth, and they'd probably be biologically incompatible with them.

Would we feel at home on Mars by that point? I don't think so. After hundreds of thousands of years of evolution on Earth, followed by hundreds of thousands of years of evolution on Mars, we will have evolved too far away from our original selves (as we were on Eden). We wouldn't feel any more at home on Mars than we currently do on Earth. We should feel a lot more comfortable though – and hopefully we would feel much happier as a result.

Of course, the many evolutionary changes we've undergone since we arrived on Earth (and which we'll undergo on Mars in the future) mean we wouldn't feel at home on Eden either, if we returned there. Nor would we feel at home anywhere else in the universe, for that matter.

If any humans remained on Eden when our ancestors were brought to Earth hundreds of thousands of years ago, *they* will have continued evolving too. It's unlikely that we would resemble them or be biologically compatible with them. We might not even recognize them as belonging to our own species or genus.

- **Our innate need to worship deities**

Every civilization on Earth seems to worship one or more deities. No one knows why this is. My feeling is that we either mistook the aliens who brought us here for gods, or the aliens wanted us to believe that's what they were. They might even have hard-wired it into our brains, ensuring it would be passed down from generation to generation.

But the aliens we mistook for gods might not be the ones that brought us here. The "god" aliens might have visited us later, and they may have given us knowledge that allowed us to advance. The aliens that brought us here may have preferred that we didn't have that knowledge.

Some people believe we worship deities because we're seeking answers about why we're here. The thought that there's someone watching over us, keeping us from harm and answering our prayers is also a source of comfort and reassurance for millions of people. It provides hope when things seem hopeless. It's something we can lean on when we're anxious or worried.

This is a reasonable thought. But as *every* community and culture around the globe worships some form of deity, there must be more to it than that.

I'm sure we would have worshiped deities on Mars if we'd been taken there instead of the Earth. And those deities may have been the same ones that we worship here on Earth.

When our descendants migrate to Mars thousands of years from now, one of the first things they'll do will be to build churches, cathedrals, chapels, shrines, and mosques, and worship the same deities as they did on Earth.

But none of this explains why there are atheists and agnostics, or why their number waxes and wanes from one generation to the next.

Nor do we know whether any deities actually exist.

• We can mend ourselves (and other animals)

No other species on Earth can do this, but we'll have to consider it an anomaly rather than evidence that we're from another planet. It's almost certainly linked to our intelligence and dexterity. At least we're compassionate as well as violent.

Some people believe we were brought to this planet to help heal the other animals. While that's a nice thought, we've driven more species to extinction than we've helped, so this theory doesn't stand up to close scrutiny.

- **We go against our natural behavior**

Anthropologists believe that some of the last remaining people to retain natural human behavior are the Bushmen of the Kalahari in southern Africa. Just about every other civilization on Earth has rejected our "natural" way of life.

> It's worth noting that the Kalahari Bushmen's life expectancy is around 45 to 50 years. Earlier human civilizations that lived "natural" lifestyles rarely made it out of their thirties.

We much prefer our modern lifestyles, even though we know it damages our health. Many of us eat the wrong kinds of food – and far too much of it; we reject any form of exercise; we ruin our health through smoking, abusing drugs, and drinking excessive amounts of alcohol; and so on. No other species on Earth knowingly and deliberately engages in behavior that damages them.

Many people believe this behavior is linked to our intelligence and technological development. Yet the opposite seems to be true; the *least* intelligent among us are the most likely to indulge in these damaging activities (or lack of activity when it comes to exercising).

> There are exceptions, of course. Many highly intelligent, talented and wealthy people have ruined their lives through drink, drugs, obesity, and so on.

> Despite this reckless behavior and overindulgence, we still live significantly longer than the Kalahari Bushmen, who supposedly live natural lifestyles.

I'm sure we would have followed exactly the same pattern of behavior if we'd lived on Mars or any other planet. We would have started out by following our "natural" pattern of behavior (basic survival) just as we did when we arrived on Earth. But later on, we would have adopted modern, urban lifestyles and more damaging behavior.

- **We age horribly**

As we get older, we don't just become grayer and more wrinkled, as other animals do; our appearance changes *completely*. If you compare photos of a young person, the same person in middle age, and the same person in old age, they look like different people.

> Most other animals look much the same throughout their entire lives – with the obvious exception of those that undergo metamorphosis, such as butterflies and moths.

Most anthropologists believe our "natural" lifespan is only about thirty years. The fact that we were able to double it or even triple it by changing our living conditions suggests that our so-called natural lifestyles (as lived by people like the Bushmen of the Kalahari) were far from optimal. If we take animals out of the wild and put them in what we think are their optimal living conditions, their lifespans usually increase *slightly*, but they don't double or triple as ours did. This is a further indication that we're markedly different from every other animal on Earth, and we're ill-suited to this environment.

As we'll see in the next section, though, it seems that we're hard-wired to go through the menopause and become grandparents. This suggests that our natural lifespan is far longer than the anthropologists believe. What they refer to as our "natural" lifestyle might be nothing of the kind. Our "survival" lifestyle when we first arrived on Earth certainly wasn't optimal – and I'd say it wasn't our natural lifestyle either.

Our living conditions on Eden (and on Mars if we made it similar to Eden) *should* be optimal. We can therefore assume that our lifespan on those planets would be at least as long as it is on Earth today. But it might be much longer.

> If you've read *Humans Are Not From Earth*, you might remember the section about Methuselah and other people who reportedly lived for hundreds of years.

I wonder whether our appearance would change so dramatically as we aged on Eden and Mars, compared with the way we age on Earth?

Is our aged appearance a factor of living on Earth, in an environment that doesn't suit us? Could it be connected with the genetic manipulation the aliens carried out when they brought us here? Or is the change in our appearance entirely natural?

Such a dramatic change is rare in other species on Earth, but it might be common on Eden.

- **The menopause and grandparenting**

When we had much shorter lifespans, women didn't live long enough for the menopause to occur. That meant they could bear children throughout their entire (short) lives – just as other animals can on Earth.

But once we started living longer, something strange happened. When women reached around fifty years of age, their monthly menstrual cycle ceased, they could no longer have children, and they switched from "parent" mode to "grandparent" mode.

This doesn't happen in any other animals on Earth, except for two species of whale – pilot whales and orcas.

> It's interesting that these are both *marine* mammals.

The majority of women go through the menopause at around the same age, and they experience it in the same way. This suggests that it's deliberate, and an inherent part of our design. It also suggests we were meant to live long enough to experience it.

> Once again, we have to wonder whether the Kalahari Bushmen's lifestyle is as natural as anthropologists think it is. On average, *none* of their people live long enough to go through the menopause, and there are very few grandparents in their society.

Perhaps most of us are now living the lifestyle we were meant to. When we first arrived on Earth, we had no tools, technology or medical facilities at our disposal. Our memories may have been erased and our thinking impaired, so we had no way of even conceiving of such things. We had no idea how we were supposed to live. So we did what the native hominins did, and eked out a living as best we could. But, given the evidence of the menopause, this was clearly not our natural lifestyle, and our life expectancy was significantly lower than it should have been.

> One has to wonder why groups such as the Kalahari Bushmen got left behind, bearing in mind they have much shorter lifespans than everyone else. Was it their deliberate choice? Perhaps the aliens who brought us here (or a different race of aliens that gave us the knowledge to advance) wanted a control group they could compare us with.

- **Savant-like skills**

A few people in the world exhibit extraordinary skills in things like mathematics, art, music, and memory. However, they tend to have significant mental impairment in other areas, and often lack social skills.

Researchers have found that by stimulating the left temporal lobe of our brains with magnetism, they can induce these same savant-like skills in people without mental impairment. Some of their experiments even appear to have induced telepathy.

Their experiments have shown that almost everyone has these skills. But most of us can't use them because the connections in our brains seem to be broken. One day, it might become possible to reconnect them. This could trigger another massive leap forward in our technological progress.

This ties in with the theory that the aliens erased our memories and disabled, degraded or suppressed certain parts of our brains when they brought us here. They may have intended keeping us permanently in the Stone Age, living what we *thought* were natural lives, but which were actually nothing of the kind.

Would they have done this if they'd taken us to Mars instead? I believe they would, because they needed to prevent us from threatening other alien civilizations.

But why didn't they do those things when we lived on Eden? Why did they go to the trouble of finding us somewhere else to live, and adapt us to help us survive here, when they could have simply left us where we were?

The most likely reason is that Eden was no longer capable of supporting us. Perhaps we had damaged it to such an extent that we could no longer live there. Or perhaps the environment had collapsed in the same way that Mars's did. After all, we suspect that Eden is smaller and older than the Earth. Its outer core may have cooled and become viscous, or even solid, and its magnetic field may have failed.

Alternatively, the aliens might have wanted to remove us from our technologically advanced world for some reason. There might have been no opportunity to do that on Eden. Other intelligent species might have opposed their idea, prevented them from doing it, or undone their work and restored our memories and technologies. But if the aliens took us to another planet, outside of their jurisdiction, they would have been able to do whatever they wanted to us.

We'll look at some of the possible reasons for this later in the book.

• Massive technological leaps

Isn't it strange that we spent nearly 400,000 years using nothing but rocks, sticks and (later) spears as our only tools, and then, apparently out of nowhere, created the entire modern world in the space of just 12,000 years?

We can pinpoint at least four periods of progress that many people believe were too rapid to have occurred naturally:

1. Around 12,000 years ago when we developed agriculture and established permanent settlements.

2. Around 5,000 years ago when the Egyptian, Roman, Greek and Chinese civilizations were founded.

3. The industrial revolution, which began in the mid-eighteenth century.

4. The age of space and technology that many people believe began in 1947 and stemmed from the reported UFO crash near Roswell, New Mexico, USA, and the technologies that were discovered on board when engineers analyzed the wreckage.

There may have been a further period of rapid progress around 8,000 years ago, when many researchers believe the many Neolithic structures around the world were created. These may have included some of the early pyramids, the

Great Sphinx in Egypt, and perhaps Stonehenge in England, along with similar structures. However, there's considerable debate about how old these structures really are.

> There's a convincing theory that the Roswell incident was actually a U.S. Military experiment that went badly wrong. Labeling it a "UFO crash" – and then denying it – enabled the military leaders to cover up the dreadful events that had *really* taken place. Nevertheless, the events surrounding this incident do seem to be connected with a massive leap in our technological progress and the beginning of the space age.

There are a several possible explanations as to how these great leaps in our progress may have happened.

- They could have occurred naturally, and our progress may have proceeded as a series of rapid leaps and plateaus. But we don't know what triggered the first leap that occurred around 12,000 years ago after around 400,000 years of plateau.

- The aliens who brought us here may have removed the limiting blocks they placed on our memories, thought processes, and other skills. This seems highly unlikely though, as we might have then developed into a species that could have threatened them. However, it may be the case that the planet, solar system and galaxy they've brought us to are so remote from other civilizations that there's no chance of that ever happening.

- A more plausible reason is that the limiting blocks may have been removed by a *different* race of aliens. We don't know why they might have done this.

  They might have been enemies of the race that brought us here, and they removed the blocks to spite them. It may have been a coded warning to them, perhaps demonstrating that they know we exist, that they can reach us, and that they can manipulate us.

  Or they may have discovered that we were created as violent soldiers, and they may be planning to use us to fight a war against the aliens that brought us here. We'll look at this in more detail in the next chapter.

- Alternatively, aliens from somewhere else in the universe may have discovered the Earth, found that it had life, realized we had the potential to advance, and given us a "nudge" to help us develop. They might not have realized we were once a highly advanced race and we retained information about that period in our genetic memories. Once they removed the blocks, we were able to access that information and rebuild the civilizations and technologies we had once had.

- Or perhaps several different alien races have visited the Earth over the last 12,000 years, and they've each given us a little nudge – or passed on technological information – to help us advance.

We'll look at this in more detail in the next book in the series.

- **Evolutionary design flaws**

There's a long list of problems with our basic design – we'll look at some of them in a moment. If we had *really* evolved on Earth, it's crazy that after billions of years of evolution these problems still exist, especially as some of them are bizarre and ridiculous.

The fact that they haven't been resolved suggests that:

- we evolved somewhere else

- or we evolved far more recently than we've been led to believe – and not from any of the Earth's native species

- or someone has interfered with our design – and they made mistakes.

The flaws include:

- Our pharynx, which we use for both breathing and for ingesting food and drink. Our airway can easily become obstructed with food matter or saliva, leading to thousands of deaths each year.

- Our inability to biosynthesize vitamin C. We have the gene for synthesizing vitamin C, but it's broken. We can only get it from our food. If foods that are rich in vitamin C become scarce, or we choose not to eat them (as many cultures do) our immune systems become weakened.

- The proximity of our genitals and rectum. This is aesthetically displeasing and unhygienic, and it's a leading cause of urinary tract infections in women. Thousands of women have been rendered infertile or even died from this.

- Our multi-function genitals. This is another example of poor design. Both men and women can contract urinary tract infections and sexually transmitted diseases because of this multi-function arrangement.

- The prostate gland. It's prone to swelling and can block the flow of urine in older men. If the urinary tract was routed around the prostate gland and connected to it via ducts, rather than running right through the middle of it, it would make much more sense.

- Our overloaded lower backs. Millions of people suffer fractured vertebrae each year, and millions more are forced to take time off work because of back problems. There's no good reason why our backs couldn't be significantly stronger and better designed. The Neanderthals (supposedly our closest relatives on Earth) had excellent back health – and they were better adapted to life on Earth in many other ways.

- Our weak knees and hips. Anthropologists say the main cause of these problems is our bipedalism. But large birds such as ostriches, emus and cassowaries, which are at least as heavy as us, have carried their weight on two legs since the Jurassic period. There were bipedal dinosaurs, including the mighty *Tyrannosaurus rex*. And there are other highly successful bipedal species including kangaroos and wallabies that don't have the problems we do.

- Our birth canal. It's unreasonably narrow compared with the size of our babies' heads. This leads to prolonged labor, significant pain, and serious health risks to both mother and baby.

- Our overly complicated feet. These may have made sense when our ancestors lived in trees and needed flexibility. But Earth's native hominins stopped living in trees around seven million years ago. So why do we *still* have unstable feet that twist as we walk, and arches that are prone to collapse? The answer is simple: we didn't evolve from Earth's native hominins.

- Our sinuses. These are supposed to drain fluid and mucus from our heads, but they're dreadfully inefficient. Fluid and mucus often drains the wrong way, or builds up, leading to infection, inflammation, and pain.

- Many of us get headaches and pains in our joints when the weather changes. We can conclude from this that the weather on Eden must be more stable, with no significant changes in air pressure or humidity.

- Our optic nerves. These are located in the center of our field of vision, causing blind spots at the *exact* point where perfect vision would be most useful.

- Our defective vision. Everyone's eyesight deteriorates as they get older, but many children have poor vision right from birth. In some Asian countries, such as Singapore, just about everyone wears eyeglasses or contact lenses.

- Our teeth. We only have a single set of adult teeth, and they don't grow continuously as they do in most other animals. Our teeth and gums weaken and deteriorate once we reach the age of about thirty-five, no matter how well we take care of them.

> We noted earlier that our digestive systems are the same as most herbivores'. Interestingly, most herbivores' teeth grow continuously throughout their lives. This is one of the few instances where we differ from them.

- Our fondness for the wrong foods. This suggests we originated on a planet where food was scarce and not particularly nutritious – but it was probably *really* tasty. As a result, our brains tell us to eat as much of it as we can, whenever it's available. The problem is that it's *always* available in most developed countries, and it's also highly nutritious. Resisting the urge to eat all the time takes more willpower than many of us possess, even when we know it could kill us.

- Our tribalism. We have an inherent need to belong to a tribe or clan, and we have an ingrained distrust or hatred of other tribes and clans. This would have been an important survival trait in our ancient pasts. But today it causes social problems including bitter rivalry, prejudice, racism, gang warfare, and terrorism.

- Our flawed decision-making skills. For example, we always agree with people who agree with us, we prefer things to stay the same even when something better is on offer, and we base our forecasts on past results.

> The "gambler's fallacy" is a good example of this. If a coin lands *heads* side up five times in a row, most gamblers believe it's more likely to land *tails* side up next time – and they'll bet heavily on that outcome. But the odds haven't changed: every coin toss has a 50-50 chance of landing heads side up or tails side up (if it's tossed correctly). So they also have a 50-50 chance of losing their money.

- Hyperactivity. This is another trait that would have been beneficial in the past. An active and enthusiastic hunter, gatherer, and fighter with boundless energy would have been a huge asset to an ancient tribe or clan. Five percent of people still have this trait today. But they're now considered "problems," and they're often medicated to calm them down.

> If we were to design the perfect human, we would correct all of these issues. In the future, we might be able to do just that.

In the next chapter we'll consider some of the many reasons why we may have had to leave our home planet.

4
Why We Had To Leave Our Home Planet

There are all sorts of reasons why the aliens might have removed us from our home planet, Eden, and brought us to this solar system. We'll consider some of those reasons in this chapter. Perhaps one day, the aliens might tell us the *real* reason. For now, all we can do is speculate and make educated guesses.

Natural disasters[4-1]

The aliens who brought us to Earth may have done so as a way of preserving our species, because Eden may have become incapable of supporting us. But why might that have happened?

Loss of the magnetosphere

We've already seen that Eden is probably smaller than the Earth, but perhaps a little larger than Mars. The problem with small planets is that their mantels and cores cool faster than they do on larger planets. When they reach a certain temperature, the molten liquid rock in the outer core becomes viscous, and the inner core is no longer able to rotate freely.

> Rather than rotating freely, it's more likely that the inner core remains fairly stationary while the rest of planet rotates around it.

The inner core begins "sticking" to the outer core and gets dragged along with it, reducing the dynamo effect that generates the planet's magnetic field. As the outer core cools further, it becomes even more viscous and the adhesion grows stronger, so the magnetic field weakens further. Eventually, the inner core rotates at the same speed as the rest of the planet, the dynamo effect ceases, and the magnetic field disappears. This is what happened on Mars, and it might well have happened on Eden too, as it's likely to be much older than the planets in *this* solar system*.

> Interestingly, Venus lost its magnetic field for a different reason. It's roughly the same size as the Earth, so it shouldn't have cooled down as quickly as Mars did. Instead, it almost certainly lost its magnetic field because of its extremely slow speed of rotation and the consequent lack of internal thermal convection.
>
> Some researchers believe Venus might not have a solid core. Or, like Mars, its outer core may have already cooled and solidified. But these things are yet to be investigated.

As we've already seen, when a planet orbiting a main-sequence star loses its magnetic field, it becomes uninhabitable. Its atmosphere is eroded by the solar wind, the greenhouse effect ceases as the insulating layers of gas are lost, the surface freezes, rainfall ceases, the surface water evaporates and photodecomposes, the surface becomes a desert, and the planet is also exposed to the full – and lethal – effects of solar and cosmic radiation and solar flares. The only way to make the planet habitable again is to restore or replace its magnetic field.

> Restoring or replacing a planet's magnetic field *is* feasible – though it's way beyond our current level of technology. We'll need to develop a way of doing it if we ever want to fully terraform Mars. We'll look at two potential methods in Chapter 9.

> Venus is the odd planet out again here, because losing its magnetic field had the opposite effect. As it lost its surface water, its atmosphere became denser, the greenhouse effect multiplied, and its surface heated up. But the result was the same: it became uninhabitable.

*How we know Eden is older than the planets in this solar system

Archaeologists have found several *extremely* ancient human artifacts on Earth. Some of them date back millions of years or even tens of millions of years, predating the earliest hominins. We looked at some of these artifacts in *Humans Are Not From Earth*. Most mainstream scientists refuse to even look at them, because they're so far outside the official timeline that they consider them "impossible." Nevertheless, they *do* exist, their ages have been confirmed, and there's no question that they were created by modern humans – long before we are supposed to have existed.

We can make several deductions from these artifacts. For example, they suggest that at least a few humans were brought to Earth millions of years before we could have evolved here. Those people can't have survived for very long, but the artifacts they left behind prove they were here. As we can prove we existed long before the other hominins evolved on Earth, that means we must have come from an older planet.

Life appeared on Earth pretty much as soon as the planet cooled enough to support it. If we assume the same thing happened on Eden, and evolution occurred there at a similar rate, then Eden must be millions of years older than any of the planets in this solar system.

But, as we'll see later, it could be up to a *billion* years older.

All of this assumes that the Theory of Evolution is correct. But it probably isn't. For example, there's little evidence in the Earth's fossil record of any intermediate species, where one species evolves into another.

There are millions of *separate* species, and we can see evolutionary changes *within* those species (via natural selection), so *that* part of the Theory is correct. But there's no evidence that individual species ever evolve into *different* species.

Again, most mainstream scientists refuse to believe this – even though the evidence (or lack of it) is right in front of their eyes.

A drab little bee might evolve into a more colorful one with bigger wings, for example, but it will still be the same species. A bumblebee can never evolve into a honey bee because it's a different species. And it certainly can't evolve into something like a dragonfly – even if you wait for billions if years.

One of the leading alternative theories is known as Intelligent Design. This is the theory that each species was individually and separately created by some form of intelligence, rather than by an undirected process such as natural selection.

Creationists attribute the intelligent design process to a deity, such as God or Allah, and they call the process creationism. Intelligent Design and Creationism are controversial subjects in the world of mainstream science, and they're commonly regarded as pseudoscience.

Advocates for Intelligent Design, Creationism and the Theory of Evolution each say theirs is the only correct theory and there's no scientific basis for any of the others.

My view is that Intelligent Design is the *only* theory that explains how the separate species are created, but the Theory of Evolution by Means of Natural Selection correctly explains how changes occur *within* those species.

Climate change

Like most things in science, there are different (and conflicting) theories about what causes climate change on Earth.

The largest group of mainstream scientists says humans caused it through their industrial activities – basically by making the atmosphere slightly darker and denser, so it traps more solar energy.

But another group says climate change is an entirely natural process[4-2] and the Earth goes through cycles of warming and cooling, depending on its orbit, solar activity, ocean currents, volcanic activity, and weather-related phenomena such as El Niño and Arctic oscillation.

In reality, the current period of climate change on Earth is almost certainly caused by a *combination* of these two things – both natural and man-made.

Climate change has occurred ever since the Earth first formed over 4.5 billion years ago. The planet seems to be going through a natural warming phase at the moment, but it's been through numerous cooling phases in the past, such as during the various ice ages[4-3]. There were times when it may have frozen over completely, such as during the Great Oxygenation Event when around two percent of the carbon dioxide in the atmosphere was replaced by oxygen.

The same processes must undoubtedly occur on Eden. And we might have been partially or wholly responsible for making it uninhabitable.

> Carbon dioxide is a greenhouse gas that acts as an insulating layer in the atmosphere. It traps solar radiation and helps keep a planet's surface warm, just like the glass in an actual greenhouse.
> Oxygen does not do this.

Atmospheric change

Something else might have altered Eden's atmosphere. For example, global warming or an asteroid strike might have caused trapped carbon dioxide to be released from the rock. This would have made the atmosphere denser and increased the greenhouse effect to make the surface warmer. The increase in carbon dioxide might also have made the air so toxic that we could no longer breathe it.

Asteroid or comet strike

Eden could have been hit by an asteroid or a comet, causing widespread destruction, crop failures, and so on. If the impact was massive enough, all life on the planet could have been wiped out, or the planet might even have been destroyed. The few humans who survived (who would have been our direct ancestors) might have been visiting or living on a neighboring planet at the time. The aliens kindly found them a new home here on Earth.

Within the next century, we might develop the technology to alter the paths of small-to-medium-sized asteroids and comets, and divert them away from the Earth – as long as we spot them early enough. But larger asteroids – or those we don't spot until it's too late – could still wipe us out.

We don't know what sort of technology we had when we lived on Eden. We were probably millions of years ahead of where we are now. But altering the path of a large asteroid or comet might still have been beyond our capabilities. It will be beyond our capabilities here on Earth too for the foreseeable future. Perhaps we'll find a way to do it in ten million years or so.

> We probably tried to find a way to redirect the asteroid or comet before it hit Eden. But we may have run out of time.

> In about one billion years' time, scientists believe we might develop the technology to move the Earth itself. But that might be too late. As we'll see in Chapter 10, we'll probably need to evacuate the planet long before then.

Loss of a moon

Eden might have been destabilized when its moon (or perhaps one of its several moons) drifted too far away. As a result, the formerly stable planet might now have no tides, rotate more slowly, wobble on its axis, experience extremes of weather, and so on. This could have caused the planet's entire ecosystem to collapse.

> Many people believe the same thing will happen on Earth in several billion years' time. The Moon is currently moving away from us at a rate of 1.48 inches (3.78 cm) per year – which is about the same rate as your fingernails grow. However, as it drifts further away, the rate of drift is slowing down because there's less gravitational interplay with the Earth. Most astronomers believe the Moon will never leave the Earth's orbit; it will eventually reach a point where the drift stops. But as the Moon will be much further away than it is now, its influence will be reduced and our tides will become smaller.

> Even if the drift continues forever, it will take the Moon so long to leave the Earth's influence that the Sun will have become a red giant by then and all life on Earth will have been wiped out. So no one will be here to see it anyway.

> Astronomers predict that the Sun will enter its red giant phase in around 5.4 billion years. Over the following two billion years, it will expand massively and consume Mercury and Venus, and perhaps the Earth too. We'll look at this in more detail later in the book – and consider some possible ways of surviving it.

> If a planet's entire ecosystem collapsed, all of its plant life would die, the carbon dioxide level would rise, and the oxygen level would plummet. If we lived on a planet where this happened, we would have to move into domes on the surface or into sealed habitats underground, and grow crops hydroponically. This is the same sort of technology that we'll use when we begin colonizing Mars. We'll take a closer look at it in Chapter 9.

Solar activity

The star that Eden orbits might have started to expand into a red giant. This process typically takes around two billion years, and some of our ancestors were almost certainly still living on

Eden as recently as 100,000 years ago, so I don't believe its star could have become be a red giant yet[4-4].

But if it's anything like the Sun, about four billion years before it becomes a red giant, it will become ten percent more luminous — with a similar increase in heat energy. This will cause a runaway greenhouse effect and turn Eden into a facsimile of Venus: a dry, lifeless desert with a dense atmosphere that can crush metal. All life will be wiped out. This process may have begun on Eden when the aliens rescued us from it.

> The same fate awaits the Earth in about 1.1 billion years. Hopefully our descendants will have migrated to Mars and other planets by then. If not, they'll have to live underground (or move the Earth into a wider orbit) if we are to survive as a species.

Coronal mass ejection

Another form of solar activity that could have wiped us out is a coronal mass ejection (also called a solar flare) from our home star. This is a massive release of plasma, and an associated magnetic field. Our Sun produces coronal mass ejections every three to five days, but they shoot out into space in random directions and only hit the Earth once every few years. They usually cause nothing more than spectacular auroras and occasional power outages. Very rarely, a large ejection can have a serious impact, disrupting the Earth's magnetosphere and taking out continent-wide electricity and communications networks. If there were no back-up systems, things like water supply systems to urban communities could fail, potentially leading to millions of deaths.

> The last coronal mass ejection of this size was in 1859. It disabled parts of the newly created U.S. telegraph network and caused electrical fires. No such ejections have occurred (yet) in the modern electronic era – though one nearly hit us in 2012.

Environmental collapse

Eden's environment could have collapsed for a different reason than any of those we looked at above.

Global pandemic

Bubonic Plague (also known as the Black Death) killed between 75 million and 200 million people in the fourteenth century[4-5] – up to half of the Earth's population at that time[4-6]. If a pandemic of that size had struck our home planet, it could have all but wiped us out.

We've already seen that the human population on Eden probably wasn't particularly large in the first place. Only a few small, isolated pockets of survivors might have remained after the pandemic, and there might not have been enough of them left to form viable breeding groups.

Famine

Our small population would have been highly susceptible to crop failure and famine. Many of the issues we've looked at in this chapter, including climate change and other natural disasters, could have caused this, and reduced our numbers to the point where we became an endangered species.

Man-made disasters

When you consider the many terrible things we've done on Earth, it's easy to imagine that we could have rendered Eden uninhabitable all by ourselves. Here are just a few of the things we might have done.

Nuclear war

A global thermonuclear war could have wiped out millions of people and blasted millions of tons of ash and debris into the atmosphere. That in turn could have caused a nuclear winter that lasted for several years, leading to crop failure and famine, an increased carbon dioxide level, a reduced oxygen level, polluted water supplies, and more.

Given our propensity for violence, it's all too easy to imagine this scenario. In fact, many researchers believe this could be one of the main reasons why we haven't (officially) found any other extraterrestrial civilizations. The researchers theorize that almost every civilization eventually destroys itself.

> This is a negative view based only on our knowledge of ourselves. I believe there are plenty of benign species that live in harmony with their planet, fellow beings, and other civilizations, and they would never destroy themselves. The reason why we don't encounter them is because we've been placed in a remote part of the universe, and they've been instructed to stay away from us.

> Of course, we are often visited by aliens and UFOs. Some of them may have been sent to monitor us and disable our nuclear weapons. But as so many of them have apparently crashed on Earth, it's likely that many of them are simply joyriders breaking the rules and thumbing their noses at authority.

> Officially, we haven't encountered any other extraterrestrial civilizations. In fact, we have, but the details are highly classified and kept secret from the public – at the aliens' request. However, we *have* officially encountered unidentified flying objects (UFOs – also known by the U.S. Military as unidentified aerial phenomena or UAPs)[4-7].

> Although there are almost certainly plenty of benign extraterrestrial races, they would be invaded by less benign races and wiped out if they were unable to defend themselves. There's an interesting theory that one of those benign races might have created us as ultraviolent warriors to help them fight (and win) a war against an invading race. We'll look at this in more detail later in the chapter.

Nuclear accident

A nuclear accident could have had dire consequences, similar to those described above. It might have taken one of the following forms:

- a weapons test that went seriously wrong

- a test-firing of a missile that was interpreted as a hostile action by another country or planet, and they retaliated

- a laboratory test that went wrong

- a large nuclear power plant that suffered a meltdown

- a nuclear-powered vehicle failure – especially if it was a large spacecraft (such as a mothership) parked in orbit

- or something else

Man-made pandemic

We looked at pandemics in the section on natural disasters. But what if the pandemic was something we'd created ourselves? There are all sorts of scenarios here. For example:

- Virologists on Eden might have tried to modify a virus, perhaps to reduce or nullify its infection rate or the seriousness of the disease it caused. Their efforts may have gone badly wrong. When they injected it into test subjects, they may have spread it from person to person and ended up infecting and killing millions of people.

- Or they might have been working on a virus that only infected other animals, but accidentally modified it in such a way that it became able to infect people.

- A technician might have spilled contaminated liquid from a test tube that he thought contained only water. He may have become infected and spread the virus to his family, who then spread it to others, and it spread around the whole planet.

> Many people believe that one of these scenarios led to the Covid-19 coronavirus pandemic on Earth. The virus is believed to have originated in Wuhan, China, which is home to the Wuhan Institute of Virology, a well-known virus research institute.[4-7a]

- The virologists might have been tasked with modifying the virus to create a biological weapon. Perhaps it was never meant to be unleashed – but somehow it was.

- Biologists might have been trying to create an artificial organism. They might have designed it to be benign, but failed to understand that some genes express themselves in more than one way. When those genes interacted, they may have made the organism toxic, malignant and highly infectious.

- Geneticists might have tried modifying something we eat. We noted earlier that food may have been scarce on Eden, and it probably lacked nutrition. They may have been working on ways of improving these issues. In doing so, they might have accidentally made the food toxic and caused millions of deaths. Or perhaps it caused genetic issues in children that rendered them weak, prone to infection, or infertile. This might have gone unnoticed until they reached adulthood more than a decade later. Subsequently, the birth rate might have plummeted, driving us to the brink of extinction.

Hi-tech terrorism

Many researchers have speculated about what might happen if a well-organized terrorist group was able to get its hands on nuclear weapons – or the components to make them. The consequences could be equally dire if they were able to get hold of (or create) chemical or biological weapons, or if they were able to release a man-made or genetically modified virus that caused a global pandemic.

> You should be in no doubt that at least one terrorist organization on Earth is currently trying to do one or more of those things. Eventually they will succeed.

Artificial intelligence turns malevolent

Futurologists often speculate about what might happen if we created a form of artificial intelligence that was able to create ever more intelligent versions of itself. We *will* undoubtedly do this one day. We must hope it doesn't decide that humans are unnecessary, an impediment to its progress, or harmful to the planet, and devise a way of destroying us.

Programmers of such systems will undoubtedly build in preventative measures so they can't cause us harm. But as the machines create new iterations of their programming code, they might remove these measures or find ways of overriding or bypassing them – especially if they were poorly implemented or the programmers made mistakes, took shortcuts, or skipped some of the testing procedures. This may have led to our demise on Eden, and the planet might now be ruled (or solely occupied) by sentient machines.

> I wonder whether the machines will attempt to conquer other planets and galaxies? As they become ever more intelligent and capable, there might be no way of stopping them.

Overpopulation

We may have found ways of increasing the yield and nutritional content of our food on Eden, just as we've done on Earth. Our population size might then have exploded, to the point where the planet became endangered and other species were threatened with extinction. We might have begun tearing down the forests to create more farmland, ripping the planet's crust open to access the minerals, releasing harmful gases into the atmosphere, and so on.

As we're doing exactly the same thing on Earth right now, the aliens who brought us here might be considering relocating us to yet another planet.

A better solution would be to relocate us to *several* different planets, perhaps placing a billion people on each one, and leaving a billion of us here on Earth. However, given enough time, we should be able to achieve this all by ourselves. So perhaps the aliens will bide their time for now and see how we get on, and only step in if they really need to.

> They might give us another helping hand and pass on some more of their advanced technological knowledge. As we saw in *Humans Are Not From Earth* – and as we'll discuss further in the next book – there's plenty of evidence that they've done this already.

Violence and destruction

We've already seen that we are inherently violent and destructive – far more so than any other species on Earth. We don't know why, but it may have led to us being banished from Eden. On the other hand, it could be the reason why we exist in the first place. We'll come back to this point later.

Global destruction

We can see examples of our destructive behavior all over the world. We strip acres of rainforest bare every day, we rip up hedgerows and trees – home to countless species of wildlife – so we can create larger areas of farmland that are easier to manage with machinery. We rip minerals from the earth; release harmful gases into the atmosphere; and dump waste nuclear material, toxic chemicals and untreated sewage into the oceans. We discard anything we no longer have a use for, carelessly dropping it on the ground, tossing it into hedges, throwing it from car windows, or throwing it into lakes, seas, rivers, canals, or other waterways. Some people do this without any thought for the consequences. They've always done it, and it's always been fine, so it will always be fine.

But it *won't* always be fine. It has *never* been fine.

As we become ever more technologically capable, this destruction is only likely to increase. For example, we might start demolishing mountains to access the minerals beneath them and create new routes. That in turn might lead to catastrophic changes in our climate, as well as destroying the habitats of endangered species.

Our oceans and waterways are now so polluted and full of plastic that they pose a serious health hazard. The plastic and other toxins, including highly poisonous heavy metals, are ingested by fish and other sea life that *we* then eat.

The result is not unexpected. More of us are becoming chronically sick, male sperm rates are decreasing, and our lifespans – which had been steadily increasing for several decades – are now reducing in some countries and static in many others.

Of course, we're also affecting the lives of every other species on the planet, as well as contributing to the change in climate – to some degree at least.

Imagine if this continued. The Earth would become no longer capable of supporting us – or any life at all.

We've been destroying the Earth at an ever-increasing rate for the last two hundred years or so. But imagine if we had done exactly the same thing on Eden, and it had been going on for *millions* of years.

Hopefully, once we were removed from Eden, it will have recovered. Or perhaps the aliens that brought us here cleaned it up once they'd gotten rid of us. I don't think they'll let us return.

Causing extinction

When the aliens brought us to Earth, they probably had no idea we would inflict such massive damage on the other species.

Within 40,000 years of our arrival, we had driven all of the native hominins to extinction. We may have integrated with them initially, or even relied upon them for help, because the aliens had dumbed down our brains and erased our memories. But the integration and support didn't last for long. We quickly gained the upper hand as our violence, greed and territorialism took over.

Since then we've caused thousands of other species to become extinct, and a million more are now threatened with extinction because of our behavior. The main problem is that we destroy their habitats for farming and other commercial reasons[4-8].

There's every reason to believe we would have caused the same problems on Eden. The aliens may have stepped in and removed us from our home planet to prevent further destruction and extinction.

Threats to other civilizations

As well as causing problems on our own planet, we might have become capable of accessing nearby areas of space, and we may have begun causing problems there too. For example, we might have begun terraforming another planet in the same solar system as Eden, perhaps with the intention of colonizing it and expanding our territory. Or we might have begun mining it for minerals and other resources.

We might not have been aware that the planet we were attempting to mine or colonize was a restricted or protected area. It might have harbored a rare form of life, and the alien civilizations in that region might have agreed to preserve it. But as we were excluded from that conversation, we knew nothing about it. So we blundered in and started hacking at it.

But we might have been even more advanced than that: not just capable of mining and terraforming *un*inhabited planets, but of invading and colonizing *inhabited* ones.

We have a history of doing this sort of thing on Earth, as people such as the Native Americans and Maori will testify. Unlike the Native Americans and Maori, however, the inhabitants of the other planets we invaded might have been able to fight back, and there could have been a massive war.

Or perhaps other races on neighboring planets witnessed our brutal attempts at colonization, and they stepped in to stop us. They not only removed us from the planet we were invading, but from our own planet too. And they banished us to a place so remote that we would never trouble anyone again.

Ultraviolent Warrior Theory

This leads us to another possibility that I hinted at earlier: that our violence was created deliberately.

There's an interesting quirk of our design that's worth noting. Our bodies (mostly) follow the conventional design of herbivores, except that we have forward-facing eyes that are generally only found in predators. This suggests we're designed to hunt other animals, but we're not designed to eat them – which sounds a bit odd. What could we be hunting, and why?

Could it be that we were created as soldiers, and the "animals" we were designed to hunt (but not eat) are enemy soldiers? That could explain where our violence comes from.

Why would the aliens have made us so violent? One possibility, which we briefly considered earlier, is that the aliens were fighting a war against another alien race. They may have been a benign species who were unable or unwilling to fight the war themselves. Or they may have been too weak or lacking in numbers. So they created us as ultraviolent soldiers to fight the war for them.

If that was the case, we seem to have done a good job. We helped them win the war, and the grateful aliens decided to keep us alive rather than exterminate us once we were no longer needed.

Of course, keeping us alive might also have been an insurance policy in case they ever needed us again. But what could they do with us in the meantime? They certainly didn't want millions of highly intelligent, ultraviolent soldiers hanging around and causing trouble – which we definitely would have done. Relocating us to a nearby planet wasn't an option: they didn't want a dangerous species like us right on their doorstep. They needed to put us somewhere well out of the way. And it had to be a place so remote from themselves and other civilizations that we would never pose a threat to them.

As we noted earlier, the aliens could have easily altered our genes to make us docile and conformant. But it seems they decided to retain our innate violence, territorialism and appetite for destruction – almost certainly because they thought they might need us again one day.

Another theory suggests that the aliens were unable to disable our violence, which remained even after our brains were impaired and our memories were erased. In fact, we may have become even more dangerous without our brainpower and memories. We may have become savages that acted on instinct and couldn't be reasoned with. The aliens' only options would have been to remove us or exterminate us. And, as we'd helped them win the war and they might need us again in the future, they'd decided not to exterminate us.

> Another factor that ties in with this theory is that nothing brings us together like a war. We have "wartime spirit." It's a time of patriotism, national pride, endeavor, and morale boosting, when everyone gets on with the job and works well together. We're *really good* at wars. The problem is that when we have no wars to fight, we turn on each other instead.

Prison Planet Theory

Another theory that ties in with our inherent violence is the idea that the Earth is a prison planet. We're violent troublemakers and invaders, and we have to be kept isolated from other civilizations. It's a familiar theme.

We've already noted that the galaxy we now live in is in one of the remotest parts of the universe, and our solar system is, in turn, in one of the remotest parts of that galaxy.

The aliens are also monitoring and restricting our activities. We've frequently observed UFOs near our nuclear facilities, and some military facilities – including nuclear missile launch sites – have been temporarily disabled. Some researchers believe *every* civil and military nuclear facility on Earth has received at least one UFO visit.

It's highly unlikely that the aliens will allow us to venture very far into space. In 2019, NASA's *Voyager 2* spacecraft detected an 89,000°F (50,000°C) wall of plasma surrounding the solar system[4-9]. Although *Voyager 2* passed through it unscathed, humans might not be able to, and it may represent the limit of our travels.

If this is confirmed, then it's bad news for our species' long-term survival.

> Although we speculate that there might be life on other planets and moons in this solar system, if the prison planet theory is correct, then there probably isn't any life there. (Or, if there is, it probably isn't terribly important.)
>
> The extraterrestrials must have known we would eventually visit those places, contaminate them, and even try to terraform and colonize them. If we did that, we would have a detrimental effect on any life-forms that were already there, and we might even destroy them. If the life-forms (or remains of former life-forms) on those planets or moons were in any way significant, the aliens would prevent us from going there. And it's highly unlikely that they would have brought us here in the first place.
>
> One has to wonder, in that case, whether life on *Earth* is in any way significant. Perhaps it isn't — from the aliens' point of view.

Extraterrestrial causes

There's a small possibility that we might not have caused any problems on Eden at all; it might be the *aliens* who are the troublemakers.

They may have invaded our home planet, kicked us off of it, and set about mining and asset-stripping it.

Being the violent species that we are, I'm sure we didn't go down without a fight. But the aliens would have had vastly superior technology, and we wouldn't have stood a chance. Another alien race might have rescued us and brought us here so the hostile aliens wouldn't be able to find us.

Another possibility is that the aliens felt our population was becoming too large, and they decided to do something about it. They might have unleashed a virus and caused a pandemic that all but destroyed us. They might have intended to wipe out every last one of us. But somehow a few of us survived and another species rescued us.

As we noted earlier, the only survivors might have been on a neighboring planet at the time. They might even have been prisoners there. When they were released, the aliens brought them to Earth, so that our species didn't become extinct.

> There might be a galactic or universal rule that says no species should knowingly and deliberately cause the extinction of another. I certainly hope so.

The opposite scenario is also a possibility: rather than our population becoming too large, it might have dwindled to such an extent that we had become endangered. Eden might have become no longer capable of supporting us, and the aliens found us a new home. They might even have established it as a breeding colony in an effort to boost our numbers and save us from extinction.

One last possibility is that Eden might simply have been in the wrong place. The British science fiction/comedy writer Douglas Adams might have been closer to the truth than he realized when he said in his novel *The Hitchhiker's Guide to the Galaxy* that the Earth was in the way of a hypergalactic bypass and had to be demolished. Eden might have been blocking or impeding signals between alien planets, blocking a view of something in space, preventing access to a wormhole, or generally getting in the way of things. Whatever the reason, it had to go, and we needed to be rehoused.

In the next chapter we'll take a detailed look at planet Mars. We'll examine its geology and other factors, and learn why it would make it a much more suitable home for us than the Earth.

It's important to note that our survival as a species depends on our ability to colonize Mars. The Earth will become uninhabitable before too long – and we don't have as long as you might think.

5
Mars Past And Present

In this chapter we'll take a detailed look at planet Mars and its geology, both today and in the past. We'll also compare it with the Earth and see why it would make a much better home for us – assuming, of course, that Mars was habitable.

When the aliens that brought us here first came across Mars, it would have seemed a much better home for us than the Earth. We can see from the evidence in our own bodies that Mars would have closely resembled our original home planet, Eden. At that time, Mars would have had a dense atmosphere and plenty of liquid water on its surface. It would *not* have had breathable air, but the aliens would have known that if it followed the pattern of other planets they'd studied, and it developed a dense covering of plant life, the air *would* eventually become breathable. They might even have addressed this issue themselves by beginning the process of terraforming it.

> They might also have terraformed the Earth at the same time. This would have given them a backup planet in case terraforming Mars failed.

If we look at the Mars of the past, and compare it with the way it is today, it's clear that the aliens must have come across it more than a billion years ago, and perhaps as long as three or four billion years ago.

Their (unmanned) spacecraft might have visited each of the planets in this solar system, taken measurements and observations, and sent the data back to the aliens' home planet. When they needed a new home for us, they might have searched their database for a planet that closely resembled Eden but was extremely remote from other civilizations. Mars may have been at the top of their list.

Hundreds of thousands of years later, the aliens might have sent spacecraft back to Mars to check it was evolving as they expected – which it was. And then they might have begun preparing to ship us *en masse* to our new home.

The problem, of course, is that Mars was in a state of terminal decline. The aliens' later expeditions will have identified this. The planet was turning (very slowly) into a frozen, barren, uninhabitable wasteland.

Mars would still have been habitable at this point, though. It retained enough of its atmosphere and surface water until at least a billion years ago, and perhaps until just a few million years ago. But eventually the atmosphere would be stripped away by the solar wind, some of the surface water evaporated and was swept away into space, and the remainder disappeared underground.

Why it went wrong

As we saw earlier, the main thing that caused these problems was that Mars's lost its protective magnetosphere. As a result, it was exposed to the full effects of the solar wind – a stream of charged particles that leave the Sun's corona at up to 600 miles per second. The particles exerted a pressure that eroded the atmosphere. It has managed to retain a very thin atmosphere, but the air pressure is only one percent of the Earth's – little more than the vacuum of space.

We also know that Mars was struck by at least one asteroid around four billion years ago, and it may have been struck by up to twenty of them. One of them was so large that it may have melted half of the planet's surface[5-1].

> Asteroids are huge. They can be hundreds of miles across – the size of a small planet. If one of them hits a planet, it causes massive global destruction and extinction. Fortunately, it doesn't happen too often.
>
> Meteors are smaller than asteroids. Large ones can be several miles across, but they're fairly rare. Smaller ones can be about the size of a car or a van. But most meteors are tiny, ranging in size from a grain of sand up to a pebble, or occasionally a large boulder.
>
> If a meteor hits a planet, it's known as a meteorite. Meteorites are common: about six thousand hit the Earth each year – an average of around seventeen each day[5-2]. Most of them are so small that they burn up in the atmosphere or break up

> when they hit the surface, leaving no trace behind. Many of them fall into the ocean or in uninhabited areas, where they go completely unnoticed. Only a handful fall in inhabited areas or are large enough to be tracked, recovered and studied.
>
> Only one person is known to have been killed by a meteorite – in Sulaymaniyah, Iraq in 1888[5-3]. But more than a thousand people were injured when one exploded in the air over Chelyabinsk, Russia in 2013.

Day, years and seasons on Mars

Martian days are about forty-three minutes longer than they are on Earth. As we've noted, this is a good thing as far as we're concerned, because days on Earth feel too short.

Days on Eden are probably around twenty-five hours, as this ties in with our body clocks. Mars is reasonably close to that.

Years are much longer on Mars though: 687 Earth days compared with the Earth's 365.25 days. A year on Mars would last for more than twenty-two months. This wouldn't matter, except for the effect it would have on the seasons.

Mars has seasons, just like the Earth. If Mars was habitable (or terraformed), its seasons would have a familiar feel to them, as both the Earth and Mars are tilted on their axes to a similar degree. The big difference is that the seasons on Mars would each last for five and a half months, compared with three months on Earth.

> Some researchers believe we would not be able to tolerate seasons that lasted that long. We'll look at a possible way of overcoming this issue later.

Land mass, density and gravity

Mars is about half the diameter of the Earth, and it's less dense. Its volume is fifteen percent of the Earth's, while its mass is just eleven percent. Its gravity (as measured on the surface) is about thirty-eight percent of the Earth's. A fully grown adult would weight about the same as a ten-year-old girl does on Earth. I think we can all agree that we'd be happy with that. We'd feel lighter and more agile, we wouldn't need heavy skeletons and musculature as we do on Earth, and we wouldn't get anywhere near as many problems with our backs.

Unlike the Earth, there's no surface water on Mars any more – it's all dry land. In fact, as so much of the Earth's surface is covered by water, there's only *slightly* less dry land on Mars than there is here. That wouldn't have been the case in the past, of course, and nor will it be in the future if we manage to restore the surface water and water cycle.

We know Mars once had a massive northern ocean, and the whole planet was covered in seas and lakes – though we don't know whether they were liquid or frozen. So it would have had far less dry land than it does today.

> Researchers have found evidence of a huge meteorite strike in Mars's northern ocean about 3.5 billion years ago. It created a mega-tsunami 1,000 feet (309 meters) high[5-4], so the ocean must have still existed at that time. This was at least 500 million years after Mars began losing its water and

> atmosphere. In fact, the ocean almost certainly continued to exist for two billion years or more after that. But it would have become gradually smaller and shallower.

> Mainstream scientists theorize that Mars began losing its water about 3.8 billion years ago, suggesting that the oceans began drying up then. But recent evidence from orbiting spacecraft indicates that *new* oceans began forming 3.6 billion years ago. They were probably created when some of the ice at the poles was melted by volcanic activity[5-5].

Geological activity

Almost all geological activity on Mars ceased around four billion years ago — just 500 million years after the planet formed[5-6]. Research suggests Mars once had two tectonic plates, which could have caused earthquakes as they moved against each other. But the crust is now a single, solid mass that barely moves, and there are no rifts or subduction zones. That means there are no large earthquakes. But there are occasional small tremors (also known as marsquakes). These are probably caused by liquid and frozen water moving around in underground fissures and tunnels due to the tidal effects of Mars's two moons, Phobos and Deimos.

Mars once had several active volcanoes, but they're all extinct now. The crust and mantel have become too thick for molten rock to break through from the core. The best-known Martian volcano is Olympus Mons, the largest-known volcano in the solar system. It's truly massive, standing more than 13.5 miles high. In comparison, Mount Everest, the tallest mountain on Earth, stands just 5.5 miles high.

> The reason why the volcanoes on Mars are so large is because it doesn't have any tectonic plates. The crust doesn't move around as it does on Earth, so each time a volcano erupts, it adds a new layer of lava on top of the previous layers.

> On Earth, the crust shifts slightly between eruptions, so we tend to get chains of smaller volcanoes rather than single massive ones. Examples include the Hawaiian Islands (which formed when a tectonic plate moved over a hot spot in the magma), and places like the Andes, the Aleutian Islands, and the Philippines (which formed where two tectonic plates met).
>
> There are a few thinner or weaker spots *within* the Earth's tectonic plates, as well as hot spots in the magma below them, so some volcanoes *do* remain in the same place. But as they aren't near the edges of the plates, and the magma has to force its way up through fissures in the crust, less of it reaches the surface. This means that single volcanoes, such as Mount Etna, Mount Fuji, Mount Kilimanjaro, and Mount St. Helens, never grow as large as the ones on Mars.

As we've seen, we have no natural ability to detect earthquakes and other natural disasters, unlike the other animals on Earth. This suggests that the level of geological activity on Eden must be practically non-existent. Most, if not all, activity must have ceased long before we evolved. This provides further evidence that Eden must be much older than the planets in this solar system.

Water

There are all sorts of theories about where Mars's surface water went. Many mainstream scientists believe it evaporated and photodecomposed when Mars lost its atmosphere, and it was then carried away into space by the solar wind. But the latest data from orbiting spacecraft suggests that a significant amount of the water that once existed on Mars is still there.

There are huge reserves of it locked up in the ice caps. The spacecraft have also discovered a massive frozen reservoir beneath the northern ice cap, as well as multiple frozen lakes beneath the sand dunes and between layers of sand and rock.

The latest research suggests that an increase in volcanic activity – possibly caused by the asteroid strikes four billion years ago – made the basalt rock twenty-five percent more porous than normal. It may have absorbed an entire ocean up to two miles deep[5-7].

If all of the water on or near the surface of Mars melted today, it would cover the entire planet to a depth of about 115 feet (35 meters). But if *all* of the water on Mars melted, including the vast reserves that lie deeper inside, it would cover the entire planet to a depth of at least a mile – and perhaps a lot more.

Some of the water *will* have evaporated and photo-decomposed, of course, but we don't know how much. It's certainly a lot less than our scientists once believed.

Martian rovers have examined the water that lies just below the surface of Mars, and they've found that it might be too salty and acidic to support life. However, some extremophile bacteria from Earth almost certainly *could* survive. And Mars may have evolved its own strains of extremophile bacteria and other life-forms that could not only survive but thrive in those conditions.

Even if no life-forms could tolerate those conditions, there's plenty of fresh water elsewhere on the planet. The ice at the poles is reasonably pure, and there are large frozen lakes and reservoirs deep underground, where they would be shielded from radiation and solar flares. If some of these are *freshwater* lakes and reservoirs, and if some of them are liquid – as I'm certain they must be – there's no reason why life couldn't flourish there today, at least in microbial form.

The fact that there's so much water on Mars today is useful, as we'll need it when we terraform the planet. We'll look at this in more detail in Chapter 9.

The water would need to be processed to make it drinkable, or to make it suitable for irrigating crops, but there seems to be no shortage of it. And we've already developed the necessary treatment and desalination processes here on Earth.

> As the water on Mars is so acidic, and it once covered much of the surface, some researchers believe it may have destroyed any evidence of past life. Alberto Fairén of Cornell University in the USA said, "When clays are exposed to acidic

> fluids, the layers collapse and the organic matter can't be preserved. They are destroyed."
> Searching for organic compounds on Mars will be extremely difficult[5-8].

Soil [5-9]

Mars's soil consists of a thick covering of fine, dry, talc-like iron oxide dust that constantly blows around. This, along with the iron-rich rocks, gives the planet its famous rust-red color. The soil is slightly alkaline (pH 7.7) – unlike the water which is acidic. The soil also contains many of the same elements and nutrients found on Earth, including chlorine, magnesium, potassium and sodium[5-10]. There's a problem with the chlorine though, because there's a lot of it, and some of it has bonded with oxygen molecules to form an oxidizing salt called perchlorate. The high concentration of perchlorate (between 0.5 and 1.0 percent) makes the soil and dust highly toxic.

> Some microbes on Earth use perchlorate as a source of energy. If any microbes have evolved on Mars, they might do the same.

If we wanted to grow crops from Earth in the Martian soil, we would first need to remove the perchlorate, and add water. Fortunately, this would be a relatively simple process. Perchlorate dissolves in water, so Martian farmers would just need to rinse it out. The perchlorate could then be recovered from the water and used for fuel or to produce oxygen, and the water could be reused[5-11].

This, along with most of the other terraforming processes, could be easily mechanized. We could ship the necessary machines to Mars, set them to work, and leave them to it. Once they'd processed a large enough area, other machines could plant seeds, harvest the crops, rework the soil, and plant more seeds. By the time the first migrants arrived from Earth, an ample supply of food would be waiting for them.

Another option would be to use some of the microbes from Earth that use perchlorate for energy. We could ship billions of them to Mars, spread them over the surface, and wait until they'd absorbed enough perchlorate to make the soil safe.

In the meantime, anyone visiting Mars would need to avoid getting perchlorate on themselves, or letting any penetrate their spacesuits, because it would damage their thyroid glands and could prove fatal. In fact, Martian astronauts would need to take the same precautions as people who work in uranium mines on Earth. Fortunately we already have the equipment and processes to deal with this.

Beneath Mars's soil lies a crust composed mostly of basalt. Its thickness varies between six miles and thirty miles.

Radiation

When the aliens first discovered Mars (and the Earth), radiation would not have been an issue. Both planets were shielded by their magnetospheres, and they both had thick atmospheres. If they'd had breathable air, we could have lived on either of them. The problem is that Mars's magnetosphere only lasted for around 600 million years.

The thin atmosphere that exists on Mars today prevents a surprising amount of radiation from reaching the surface. But it offers no protection from solar flares, which would be lethal to us – and most other life-forms. However, like perchlorate, they might not be lethal to *everything*. A few species of hardy bacteria, and micro-animals such as tardigrades (also known as water bears), *might* be able to survive. In experiments, these organisms have been left outside the International Space Station for years at a time, where they were exposed to the vacuum of space, solar and cosmic radiation, solar wind, and wildly fluctuating temperatures. Remarkably, a significant number of them survived[5-12]. Most of the species became inactive or dormant while they were outside the space station, but they reactivated once they were returned to normal conditions. But one species of bacteria, *Deinococcus radiodurans*, remained active, multiplied, and even evolved while it was in space. It became smaller, and formed a biofilm – a dense colony. While some of the bacteria on the outer edges of the colony died, those in the center were protected and survived[5-13].

> I don't know whether the organisms that survived were exposed to any solar flares while they were in space. If they did, the space station may have shielded them to some extent.

> Bacteria were a serious problem on the Soviet Union's *Mir* space station. Biofilm colonies grew on several controls and instruments, and could have caused widespread malfunction. They also threatened the cosmonauts' health, as they became more infectious than regular colonies.

> A mission to Mars could be jeopardized if biofilms formed and evolved on board the spacecraft. They could cause bacterial infections that could not be treated.

> Russian scientists have claimed that some of the bacteria they brought back into the International Space Station after the experiment weren't there at the start of it. They suggest that these bacteria "must have come from space." Others have said it's more likely the result of terrestrial contamination. NASA refused to comment and referred all questions about this to the Russian Space Agency[5-14].

If Mars was hit by a solar flare, most of the organisms on the surface would be killed, while most of those living underground would survive. But it would depend on where they were when the flare struck. If they were shielded by rocks or a mountain range, or they were living in a canyon or crater, even some of the organisms on the surface might survive.

With this in mind, there's no reason why extremophile bacteria, tardigrades and similar life-forms couldn't have evolved on Mars. Some of them could still live there today. Even more advanced life-forms might exist deep below the surface.

But if we are to live on Mars ourselves in the future, our first priority must be to restore (or replace) the magnetosphere. This would divert the solar wind, radiation, and energy from solar flares safely away.

The first breathable air

Planets with breathable air are rare – the Earth is the only one we know of. As far as we can tell, planets only develop breathable air if they have an abundant covering of plant life. On Earth, that happened 470 million years ago. Before that, and for around ninety percent of the Earth's existence, the air was not breathable. Or at least *we* couldn't breathe it – but anaerobic life could.

Plants are anaerobic – they don't absorb oxygen, they produce it as a byproduct of photosynthesis. When plants became abundant on Earth, the oxygen level rose significantly. But it was toxic to most living things at that time, and some researchers believe it caused a mass extinction.

> The first land plants on Earth were simple mosses and liverworts. More complex plants came much later. Researchers believe the first flowers appeared around 250 million years ago, but they might not have appeared until 140 million years ago. The oldest flowering species we know of (*Montsechia vidalii*) was found in Spain in 2015 and dates from 130 million years ago.

As the plants began absorbing carbon dioxide from the atmosphere, the level began to fall. The carbon dioxide had been acting as an insulating layer, trapping solar radiation and keeping the Earth warm. Within 35 million years of the first land plants appearing, the entire planet was plunged into an ice age[5-15].

The Great Oxygenation Event, and the ice age that followed, wiped out almost all of the anaerobic life on Earth. But they enabled the aerobic (oxygen-breathing) organisms to evolve and flourish, creating the world as we know it today.

It's unlikely that Mars ever had an oxygenation event – mainly because radiation and solar flares would have repeatedly sterilized the surface. This means that only anaerobic organisms would have been able to survive there. And in that case, life on Mars – if there ever was any – is unlikely to have evolved beyond pond slime.

> The Earth eventually recovered from the ice age that the plants caused. But Mars is further away from the Sun and receives less solar radiation. If plants had evolved there and caused an ice age, the whole planet might have remained permanently frozen. (Which it kind of is anyway.)

The average temperature on Mars today is about -80°F (-60°C), which is similar to Antarctica on Earth. No plants can survive this. But a few microbes might be able to, if they live deep underground where the rock is warmed by the mantle and outer core, and where the water is more likely to be liquid.

> Mars's outer core and mantle have cooled enough to pretty much halt the dynamo effect that generated its magnetic field. But that doesn't mean they're cold. In fact, the temperature of the outer core is around 2250°F (1230°C) and it's under enormous pressure. It's hot enough to melt most metals, including iron, nickel and sulfur[5-16].

> As this is exactly what the outer core is made of, it remains molten. But it's no longer hot enough to flow like a liquid.

Although the *average* surface temperature is about -80°F (-60°C), the *actual* temperature varies widely, from -225°F (-143°C) at the poles in winter to +95°F (+35°C) at the equator in summer. This wide range is due to the atmosphere being thin and lacking pressure, and the soil having a low thermal inertia. Neither can store much solar radiation.

> If you stood on the surface of Mars today and you were crazy enough to take your spacesuit off, you'd get a severe case of the bends. Dissolved gases in your blood would come out of solution and form bubbles, which would block the blood flow to your heart, lungs, brain and limbs. Your lungs would rupture. Your skin and body tissue would swell to dangerous levels. You'd get a severe case of hypothermia — although that would be the least of your worries. And then you would die.
>
> The good news (if you can call it that) is that your body wouldn't decompose. It would quickly freeze solid and be buried by the dust.

If we look at what little air there is on Mars (which is almost none), over ninety-five percent of it is carbon dioxide and the remainder is mostly nitrogen.

> Carbon dioxide is lethal to humans. We can't survive if the level rises above ten percent, even if there's plenty of oxygen. On Earth, the level is 0.04 percent.
>
> There's a tiny amount of oxygen on Mars – less than 0.2 percent. On Earth it's about twenty-one percent. We can tolerate a level of just fifteen percent as long as we don't move around too much. This is the level on most commercial aircraft, because it reduces the risk of fire. But if we're active, a level below nineteen percent is considered harmful.

> Almost every theory of how we could terraform Mars suggests increasing the amount of carbon dioxide in the atmosphere so it forms an insulating layer and causes the surface to heat up. This idea has several serious flaws. For example, if we grew crops on the surface, they'd absorb a significant amount of the carbon dioxide and cause the temperature to fall again – just as happened on Earth when plants evolved. Similarly, if we wanted to make the air breathable once we'd warmed the planet up, we'd have to remove most of the carbon dioxide – and the temperature would plummet.

There are better ways of terraforming Mars. For example, we could use nitrogen and methane instead of carbon dioxide. We'll look at this (and other options) in Chapter 9.

Summary

As we've seen, Mars would have suited us much better than the Earth – if only it had remained habitable. The aliens that brought us to this solar system made a good choice – except they may have overlooked the fact that Mars was in a state of terminal decline. As a result, they had to switch to Plan B at the last minute and bring us to Earth instead.

It's clear that Mars must bear a strong resemblance to our original home planet. Eden must be smaller than the Earth and orbit a similar type of star at a similar distance, meaning that the gravity, light and radiation levels will be perfect for us. The sunlight won't dazzle us or damage our skin, yet we'll receive the perfect amount of UV radiation to create vitamin D.

The lack of tectonic plates on Mars means there are no earthquakes, tsunamis, or active volcanoes. The same must be true on Eden, as we have no mechanisms for sensing such things – unlike Earth's native creatures.

One notable issue is the lack of a magnetic field on Mars. The aliens that brought us here may have been planning to restore it when their funding ran out. Eden probably has a strong magnetic field, as we have mechanisms in our brain that can sense it, but we can't sense the Earth's weaker one.

The air pressure on Eden is probably about half what it is at sea level on Earth, and the oxygen level might be slightly lower too. This would not only make breathing easier, it would also reduce the chance of fires breaking out. Again, we have no mechanisms for sensing these, and, as we saw earlier, a rise in the carbon dioxide level sends us to sleep, whereas it wakes Earth's native creatures up. When we terraform Mars, we should aim to give it the same atmospheric pressure and composition as Eden, not the Earth.

Another factor in Mars's favor is the fact that it would have had no native hominins for us to drive to extinction. The opposite was true on Earth, and the Neanderthals, Denisovans, and two or three other species almost certainly died out because of our actions.

The solar system we now live in is located far away from the center of our galaxy. And the Milky Way galaxy itself is located near the center of the largest known void in the observable universe. Both of these factors mean we're so far from other civilizations that we're unable to communicate with them – or even know whether they exist. This ties in with the prison planet theory we looked at earlier.

One thing about Mars that might *not* suit us is the fact that its seasons are nearly twice as long as they are on Earth. We don't cope well with seasons – including the ever-changing lengths of the days, the too-hot summers, and the cold, dark winters. This suggests that Eden doesn't tilt on its axis, and therefore has no seasons. Every day will be the same length, and it will have a stable climate. We would love that.

On Mars, the summers last for more than five months – which wouldn't be too bad because they wouldn't be as hot as they are on Earth. But the winters would be colder and darker than they are on Earth, and *they* would last for more than five months too. We would *hate* that. The aliens may have attempted to give us the ability to hibernate through the long winters, but it looks as if they abandoned the project. Perhaps we're so different from other species that it just wouldn't work in us.

In the next two chapters we'll take a more detailed look at the major similarities and differences between the Earth, Mars and Eden.

- Earth is the planet we currently live on.

- Mars may have been the planet we were intended to live on – and it's where we'll have to live in the future if we are to survive as a species.

- Eden is the planet we originally came from.

6
Earth, Mars, And Eden: The Differences And Similarities

The chart below shows the main geological similarities and differences between Earth, Mars and our original home planet Eden. Obviously, the figures for Eden are estimated, but they're based on what we know about our physiologies, and how we would react on a planet where the conditions suited us better.

	Earth	Mars	Eden
Gravity (% of Earth)	1	0.38	~0.6
Diameter miles (kilometers)	7,918 (12,743)	4,212 (6,779)	~5,992 (~9,643)
Circumference miles (kilometers)	24,901 (40,075)	13,263 (21,344)	~18,846 (~30,330)
Surface area square miles (square km)	197 million (510 million)	55.9 million (144.8 million)	~113 million (~292 million)
Day (Earth hours)	23' 56"	24' 39"	~25'

	Earth	Mars	Eden
Solar radiation (% of Earth)	1	0.44	~0.5
Magnetic field	Yes	No	Yes
Atmospheric pressure*			
(% of Earth)	1.0	0.01	~0.5
(millibars)	1013.25	10	~507
(psi)	14.7	0.147	~7.35
Tectonic plates	Yes	Yes, but not any more	No
Year (Earth days)	365.25	687	~690
Axial tilt	23.44°	25.19°	~0°
Age (billion years)	4.543	4.603	~4.85 – 5.8
Volcanoes	Active	Extinct	None or extinct

* Average atmospheric pressure at sea level

We'll take a detailed look at Eden in the next chapter, but first, let's have a look at the differences and similarities between Earth and Mars.

Mars's Similarity to Earth

While the Earth is closer to Venus in terms of distance, size, composition, and gravity, Mars's similarities to the Earth (and

Eden) make it a more compelling candidate for future colonization and possible terraforming. For example:

- The Martian day (known as a sol) is much closer in length to the Earth's. A sol lasts for about 24 hours and 39 minutes, compared with the Earth's 23 hours and 56 minutes.

- Mars has an axial tilt of 25.19°, which is similar to the Earth's tilt of 23.44°. So both planets have similar seasons, but the ones on Mars last nearly twice as long.

- Both the Earth and Mars have vast reserves of water. But while most of the water on Earth is liquid and lies on the surface, on Mars most of it is frozen and lies underground[6-1].

> The amount of water on Mars has been confirmed by NASA's *Mars Reconnaissance Orbiter* and *Phoenix Lander*, and the ESA's *Mars Express*[6-2].

> Venus's axial tilt is 177.36°, which means it's more-or-less upside down and its rotation is retrograde*. However, as it's only 2.64 degrees from vertical (albeit the wrong way up), the seasonal variations are negligible.
>
> Researchers believe it once had vast oceans and seas, just like the Earth and Mars. But they evaporated when the greenhouse effect caused a massive rise in the surface temperature.

> *Retrograde means that Venus rotates in the opposite direction to all of the other planets (except Uranus). Why it does this is unknown, but a massive asteroid or other planetary body probably collided with it.

The Main Differences Between Earth and Mars

Mars is a much smaller planet than the Earth, with half the radius, just over a quarter of the surface area (28.4 percent), and only one-tenth of the mass.

Mars's gravity is just over a third of the Earth's (37.5 percent). So we would feel much lighter there, and there would be a long-term reduction in our bone density and musculature as it would take less effort to support us. Some researchers believe we might also grow significantly taller.

Another effect of the lower gravity on Mars is that we would need a greater mass of air than we do on Earth to create a breathable atmosphere. For example, we would need nearly three times more oxygen and nitrogen per square foot to create the air pressure we're used to on Earth.

> As we've already seen, it would be better to give Mars an atmosphere with only *half* the pressure of the Earth's. It wouldn't press down on us so heavily, it would be easier to breathe, and it would make the process of reinstating Mars's atmosphere significantly faster and cheaper.

We don't yet know whether there would be any negative consequences to living on a lower-gravity planet like Mars. Eden almost certainly has a lower level of gravity than the Earth, but it's probably significantly higher than it is on Mars.

Astronauts who have spent many months in space have suffered a range of long-term health issues. But, of course, the level of gravity on board a space station is close to zero. The astronauts become weightless and float around inside it, and blood tends to pool in their heads.

That wouldn't happen on Mars; they'd be able to walk around normally, but they'd feel lighter, faster and more agile. Moving around would take less effort. They'd be able to walk or run faster, for longer, and travel further without becoming exhausted. And their blood would still tend to pool in their legs, just as it does on Earth, unless their hearts worked hard enough to pump it around their bodies. So I'm inclined to believe we would get along just fine on Mars, albeit with a few changes that we've already noted, such as our bone density, muscle mass, and height.

> There are ways of compensating for the lack of gravity when astronauts are on board spacecraft and space stations. For example, we could create artificial gravity by spinning the craft (or just the section that the crew occupies) to create centripetal force.

> The larger the craft, the slower the speed of rotation would need to be. To create an Earth-like level of gravity, a ring-shaped spacecraft with a radius of about 330 feet (100 meters) would need to rotate three times per minute[6-3][6-4].

But we probably wouldn't want the level of gravity on board the spacecraft to be as high as that. Sixty percent of the Earth's gravity would suit us just fine. So we might only need to rotate the craft twice per minute.

As long as the spacecraft doesn't have any windows, the people on board shouldn't notice they're rotating.

The people inside the craft would only feel the full effects of gravity if they were in the outer ring. Anyone in the central hub would be weightless. The level of gravity would gradually increase as they moved within the spokes from the hub to the outer ring. Because of the issues this could cause, it's likely that the hub and spokes would be inaccessible to most of the crew members. Only maintenance engineers might be allowed to access to them.

Science fiction depictions of spacecraft generally place the control room at the front or center of the craft. That means the pilot, navigator, and the other members of the flight crew would be weightless. It would make more sense to place them in the rotating outer ring so they would experience the artificial gravity too.

Controlling the craft would be no more difficult than if the flight crew were in the central hub. They would have viewing screens that showed images from the front of the craft, as well as rear view and other camera angles. Again, as long as they don't have windows to look out of, they shouldn't notice that their craft is rotating.

The propulsion system and main chassis wouldn't rotate with the rest of the craft. The engines would always face the same direction, and would be linked to the control room wirelessly.

Another way of generating artificial gravity is to use linear acceleration. You'll recognize this effect because it pushes you back into your seat when a vehicle accelerates. If a spacecraft accelerated at a certain, constant rate, those inside it would experience an Earth-like (or Mars-like) level of gravity.

The acceleration required to create that level of gravity in space would get us to Mars in just a few days – and in relative comfort – rather than the several months it takes us now. We will need to develop new types of propulsion system to achieve that kind of speed though.

> Artificial gravity would be much harder to create on the surface of a planet. The conventional way of doing it is to put someone at the end of a long mechanical arm and spin them. Hopefully we will develop much better systems in the future. If low gravity proves to be a major health hazard on Mars (or on the Moon or elsewhere), the colonists could spend a few hours inside the artificial gravity system to recover.

While the Earth's orbit around the Sun is almost circular (1.67 percent deviation), Mars's orbit is more elliptical (9.34 percent deviation). As a result, Mars gets significantly closer to the Sun, or further away from it, over the course of a year, affecting the amount of solar energy it receives. This will make the seasons more extreme: the northern hemisphere will become warmer in summer while the southern hemisphere will become colder in winter.

The other major difference between the two planets is that Mars's crust has become a single, solid mass. As a result, there is no longer any significant geological activity. There are no major earthquakes or volcanic eruptions, no new mountains are being formed, and so on.

This might sound like a good thing, but it isn't. It could present us with a problem in the future when we terraform Mars and restore its atmosphere. The Earth's atmosphere is maintained by outgassing from volcanoes[6-5]. As there are no active volcanoes on Mars, we would need to maintain the atmosphere ourselves. However, as we will have developed the technology to restore and alter the atmosphere by then, maintaining it shouldn't be too much of a challenge.

Another issue, that we've already seen, is that Mars has no magnetic field. That would need to be restored or replaced too.

Mars's two moons, Phobos and Deimos, are small and don't provide the stabilizing effect that the Earth's Moon does. As a result, Mars wobbles on its axis. We've also found evidence that it has occasionally tilted right onto its side, with the equator running north to south instead of east to west[6-6].

> Uranus is also tilted onto its side. Unlike Mars, though, it remains in that position permanently as far as we know.

One other important difference is that Mars is closer to the asteroid belt, which lies roughly midway between Mars and Jupiter. Based on the cratering record on Mars, researchers estimate that the risk of a one-megaton meteorite hitting its surface is five times greater than it is on Earth. Such an event could occur as often as every three years, which would be disastrous if Mars was inhabited. In the future we should develop the technology to direct the meteors away, but it's beyond our current capabilities

A related issue is that since Mars's atmosphere is so thin, even the smallest meteorites aren't cushioned and slowed, and they don't burn up as they do on Earth. They strike the surface at full size and full speed. Restoring the atmosphere should reduce their size and impact force when they hit the surface, even if we can't find a way of diverting them.

> It could take us 100,000 years or more to fully terraform Mars. I'm sure we will have found a way of redirecting the meteors by then. The simplest approach might be to shoot them with a powerful laser beam and ease them into a different trajectory, break them into smaller chunks, or disintegrate them – depending on their size, composition, and the level of threat they pose.

Another key difference is that Mars has rich deposits of calcium and other minerals. This is important because the levels on Mars will be similar to those on our original home planet Eden. The same minerals are present on Earth, but at much lower levels. As a result, we suffer from mineral deficiencies and crave things that are rich in the minerals we lack. During pregnancy, these cravings often become intensified and take interesting and unusual forms, including dirt, ashes, chalk and coal[6-7].

The other major differences between the Earth and Mars are the result of Mars having lost its magnetosphere, atmosphere, and surface water. Terraforming it should restore all of these. When that happens, most of the following differences will be eliminated:

- The surface of Mars is currently uninhabitable and inhospitable to life. A few extremophile organisms might exist there, but probably only in a dormant state. They could not have evolved there, and they wouldn't be able to move, eat, or reproduce. After terraforming, conditions on the surface should become "normalized" and life should be able to thrive there.

If any dormant microorganisms are present, they'll be reactivated.

> The exception could be bacterial species such as *Deinococcus radiodurans*. As we saw earlier, it can not only survive for years in space, but multiply and evolve. Similar species could be living on Mars.

- The freezing temperatures on Mars. As we've seen, the average surface temperature is similar to Antarctica on Earth. Nothing could grow there even if the other conditions were suitable. Another issue is that the daytime and nighttime temperature difference (the diurnal temperature variation) is much greater on Mars than it is on Earth. It's typically around 113°F (45°C), compared with a maximum of around 60°F (15.5°C) on Earth[6-8][6-9].

> The record daily temperature variation on Earth is currently 102°F (56.7°C). But this occurred during an extreme weather event when there was a sharp change in air mass.
>
> If we wanted Mars to have an Earth-like temperature, we would need to make the atmosphere much denser than the Earth's, so it trapped more solar energy. This is because Mars if so much further from the Sun. Another way of doing this would be to add energy from an external source. For example, we could place giant mirrors in orbit around Mars and reflecting sunlight onto its surface.

> As Mars currently lacks a dense atmosphere, solar panels work well because every day is sunny (except during the dust storms)[6-10]. They will become less efficient once the atmosphere is restored. But we will undoubtedly have developed better and more efficient energy systems by then. And we could always put the panels in orbit.

- The lack of surface water on Mars. There's less water on the surface of Mars than there is in the driest desert on Earth. What little water there is evaporates quickly because of the low air pressure. Terraforming Mars would partially restore its surface water. Some of the underground water might be forced to the surface too, but probably not as much as we would need. We might have to pump it to the surface, or bring it in from elsewhere. We'll come back to this in Chapter 9.

- Mars's toxic soil. As we've seen, the Martian soil and dust contain high concentrations of perchlorate. Perchlorate is a chlorine compound that's toxic to almost all known forms of life. Nothing will grow in the soil unless the perchlorate is removed[6-11].

- Mars's global dust storms. Dust storms are common on Mars, and they can blanket the whole planet for weeks at a time. The average surface temperature drops by 7°F (4°C) for several months. Solar panels can't operate, and they become coated in dust and can't operate even after the storm dissipates. The storms can also disrupt communications with the Earth and orbiting spacecraft.

> On Earth, the nearest equivalent would be a large volcanic eruption, which can throw millions of tons of ash into the atmosphere. But even the largest of these – such as Krakatoa's massive eruption in 1883 – only reduces the global temperature by around 2°F (1°C).

> A supervolcano eruption would be a different matter. It could affect the global climate for decades. The subsequent crop failures and famines could lead to a massive loss of life[6-12].
>
> Mars's global dust storms should disappear during terraforming, as the soil would become wetter and it would coagulate. The dust storms might be replaced by other storms or cyclones. But as the level of solar energy reaching Mars is significantly lower than on Earth, they shouldn't be anywhere near as dangerous.

- Mars's toxic atmosphere. As we've seen, Mars's atmosphere is so thin that it's almost non-existent. Nevertheless, it's still toxic (at least to humans). Ninety-five percent of it is carbon dioxide, and a level above ten percent is lethal to us, even if there's plenty of oxygen. But there *isn't* plenty of oxygen; the level is less than 0.4%. We would need it to be around twenty percent to survive.

  The atmosphere is also too thin to filter out ultraviolet radiation. So much of it hits the surface that it causes molecular bonds to become unstable and break down after a few hours. As a result, nothing can survive.

> Part of the terraforming process would involve replacing the carbon dioxide with an inert gas. On Earth, nitrogen fulfills this role, making up seventy-eight percent of the air we breathe. As the level of nitrogen in Mars's atmosphere is only three percent, we would probably need to import it from somewhere else. We might be able to use a lower level of nitrogen and mix it with another inert gas, but we don't yet know if this would work.

We'll explore all of these issues in much more detail in Chapter 9, which looks at various ways of terraforming Mars.

In the next chapter we'll take a closer look at the geological and biological features of our original home planet, Eden.

7

Eden, Our Original Home Planet

As we've seen, Eden must resemble Mars quite closely, and we'd feel reasonably at home on either planet. Let's consider what we can deduce about Eden.

Location

We don't know where Eden is, but we know it isn't in this solar system.

I suggested previously that Proxima Centauri b would be a good candidate, as it's the nearest Earth-like planet that orbits another star. But I no longer believe that can be the case. Proxima Centauri is a red dwarf star[7-1], and red dwarfs are significantly smaller and cooler than G-type (yellow-white) stars like the Sun[7-2]. Any habitable planets would need to orbit fairly close to them, and we have empirical evidence that Eden is further away.

> The Sun is a yellow dwarf.[7-2a]

Earlier in the book, we noted that we age (in our minds at least) in Mars years (which last for 22.6 Earth months) rather than Earth years (which last for 12 months). This suggests that Eden must orbit its star at about the same distance as Mars is from the Sun. If that star was a red dwarf, Eden would be too far outside the habitable zone and it would be frozen solid. It's more likely that Eden orbits a G-type star.

Another factor that goes against Eden orbiting Proxima Centauri is the fact that our Milky Way galaxy is near the center of the KBC void – the perfect place to put us if someone wanted us out of the way. That means we were almost certainly brought here from somewhere much further away. We *could* be here by sheer coincidence, of course. Or we could be here because we're violent, interfering, war-mongering troublemakers. I don't believe it's a coincidence.

Age

Based on the evidence we've found on Earth – including modern human artifacts that are millions of years old – we must have evolved on Eden long before the hominins evolved here. I would estimate that Eden must be at least 250 million years older than the Earth.

As we have no natural ability to sense natural disasters, the level of geological activity on Eden must be similar to the level on Mars today – virtually non-existent. That means Eden's outer mantel probably cooled and solidified before life evolved there. If that's the case, Eden could be more than a billion years older than the Earth.

I've put Eden's age range between 4.85 to 5.8 billion years, compared with 4.6 billion years for the Earth.

Size and gravity

Based on our physiologies, biologists at the University of Virginia in the USA have calculated that we'd feel far more comfortable if the level of gravity was around forty percent lower than it is on Earth.

If we assume that this is the level of gravity on Eden, and that Eden's density is roughly the same as the Earth's, it's reasonable to deduce that Eden must be about sixty percent the size of the Earth. It would therefore have a diameter of around 6,000 miles, a circumference of around 19,000 miles, and its surface area would be roughly 113 million square miles. That puts it roughly halfway between the size of the Earth and Mars.

If we evolved on a planet where the level of gravity was forty percent lower than it is on Earth, and we were still adapted to it*, we should be able to move fairly freely between Eden, Earth and Mars. We'd feel heavier and more cumbersome on Earth, but we wouldn't be incapacitated. Similarly, we'd feel lighter on Mars, but not so much lighter that we would undergo too many evolutionary changes and be unable to return to Eden. On Eden, we'd feel light and agile and be able to move around quickly and easily.

> *We are no longer adapted to Eden's lower gravity. We've spent hundreds of thousands of years on Earth, and our bodies have evolved to cope with the higher level of gravity here. We aren't fully adapted to living on Earth by any means — that might take another several million years. But we're slowly heading in that direction.

> If we spent a long time on Mars and then visited the Earth, we'd lack the bone density and musculature to move around easily. Many visitors would become incapacitated as soon as they arrived.

> When we lived on Eden, our bones may have been bird-like and hollow. In comparison, the Neanderthals and Denisovans on Earth had solid bones. We couldn't survive here with bird-like bones. Our legs would snap under our weight, and we wouldn't be able to lift anything heavy. Our bones have thickened since we arrived on Earth, but not yet to the extent that they've become completely solid.
>
> Our bones may have begun to thicken *after* we arrived on Earth. This would have been a natural evolutionary response to the higher gravity here. But if that were the case, the earliest arrivals from Eden might well have been incapacitated, as we saw above. That could explain why they didn't survive for very long.
>
> When the aliens noticed the problem, they probably spliced some of the native hominins' genes into our genome. As a result, our bones would have begun to thicken *before* most of our ancestors were brought here. So, when they arrived on Earth, they would not have been incapacitated.

Another possibility is that the aliens may have taken our ancestors to a different planet before they brought them to Earth. The gravity there might have been higher than it is on Eden, but lower than it is on Earth. Our ancestors would have been able to survive there. They would have felt heavier and more cumbersome, but they wouldn't have been incapacitated. Their bodies would have begun to adapt to the higher gravity, and this would have prepared them for them even higher gravity here on Earth. Our ancestors might have lived on this "halfway house" planet for thousands of years – and perhaps much longer.

As we've seen, Mars's gravity might be too low for us. Moving from a medium gravity planet (such as Eden) to a higher gravity planet (such as the Earth) has led to some significant changes in our bodies. This has caused us all sorts of issues, including our weak backs and difficulty giving birth. If we moved to a planet with *lower* gravity (such as Mars) the evolutionary changes in our bodies would be equally significant. But they might not be for the better.

The changes would include loss of bone density, loss of muscle mass, weaker hearts, fluid gathering in our heads and causing life-long encephalitis, and so on. These issues would sort themselves out as we evolved to cope with our new environment, but it could take as much as several million years.

> Until we became fully adapted, we would live uncomfortable and unhappy lives.
>
> The same process (in reverse) is currently underway on Earth. We're still a long way from becoming fully adapted.

Speed of rotation

We've already seen that our internal body clocks are set to a 25-hour day. The 24-hour days on Earth feel too short, and we feel rushed as we go about our business – there's never enough time to get everything done. An extra hour each day would suit us perfectly, even if we spent it asleep – we would feel more rested and more productive when we're awake.

If Eden's circumference is 19,000 miles, and if a full rotation (one day) takes 25 hours, then its speed of rotation must be 760 miles per hour, as measured at the equator.

Axial tilt

Unlike the Earth and Mars, which are tilted by 23 and 25 degrees respectively, Eden almost certainly does *not* tilt. It would therefore have no seasons. As a result, its climate will remain constant and stable throughout the year, and the length of each day will always be the same. Our body clocks would love this. We wouldn't have the problem we have here on Earth where the length of the day keeps changing with the seasons and we have to make constant adjustments.

Magnetic field

A planet without a magnetic field is incapable of supporting life, so Eden would definitely have one. As we have cells in our brains that can detect magnetic fields, it's likely that Eden's magnetic field is strong enough that we can sense it. It's therefore likely to be significantly stronger than the Earth's.

Alternatively, if our ability to detect magnetic fields was disabled or degraded by the aliens who brought us here (because they didn't want us to move around too much), Eden's magnetic field might be no greater than the Earth's.

Another alternative is that the aliens disabled the cells because they were planning to take us to Mars, which has no magnetic field. We would have lost the ability to sense magnetic fields once we reached Mars, and the cells would have become vestigial. So they may have disabled them in, favor of giving us other strengths.

Geological activity

As we have no natural ability to detect geological disasters, we must have evolved on a planet that doesn't have them. It's reasonable to assume that a lack of geological disasters means Eden has no tectonic plates, and its crust is a single, solid mass, just as it is on Mars.

That might be because it never had any tectonic plates, but it's more likely that the crust and outer mantel cooled and solidified long before we evolved. The crust would then have become a single entity with no tectonic plates, rifts or subduction zones; and any volcanoes would have become extinct.

> Only the outermost layers of the outer mantel are likely to have solidified. The underlying layers are probably still molten and viscous, becoming increasingly liquified nearer the core. The fact that Eden has a magnetic field indicates that its outer core must still be liquid. But its inner mantel probably is too.

Atmosphere

As we evolved on Eden, it must have breathable air. This means that, just as on Earth, the oxygen level is probably around twenty to twenty-one percent and the carbon dioxide level must be well below ten percent.

As we have no natural ability to detect wildfires, the oxygen level might be slightly lower than this, as the risk of fire would then be lower too. We might have had slightly less energy as a result. But as we would have felt significantly lighter and more agile in the lower gravity and lower air pressure, and we wouldn't have had the heavy bones and musculature we need on Earth, this shouldn't have restricted us in any way. We would have been fully acclimatized to the lower level in any case – just like people who live in high-altitude communities on Earth.

As carbon dioxide sends us to sleep if the level rises, it's tempting to think there might be *none* on Eden. Yet there must be *some*, because we produce it when we exhale and most plants (on Earth at least) depend on it for photosynthesis.

It may be the case that the carbon dioxide we (and the other creatures) produce on Eden is taken up by the plants almost immediately. That means there would be almost none in the atmosphere, and we would have had no reason to react to it.

Unlike planets such as Venus, Earth and Mars, Eden might not have had a carbon dioxide-rich phase during its evolution. Alternatively, most of the carbon dioxide might have been absorbed by the rocks soon after the planet formed.

There must be some sort of insulating gas in the atmosphere to keep the temperature within a reasonable range and help shield the surface from UV rays. Carbon dioxide performs this role on Earth, but on Eden it might be performed by something else, such as methane, water vapor, sulfur, nitrous oxide, or fluorocarbons.

The bulk of the atmosphere will be made up of an inert gas, such as nitrogen, just as it is on Earth. There's a small possibility that this role could be performed by something like helium or argon. But helium would tend to rise into the upper atmosphere, and argon is less abundant than nitrogen on most planets. Additionally, nitrogen is a component of chlorophyll and amino acids, so the plants could use it. The plants on Earth couldn't survive without it.

> As the level of carbon dioxide on Eden might be significantly lower than it is on Earth, the plants there might depend more on nitrogen and less on photosynthesis for their energy. So they might not be green – which is the color of chlorophyll. They might be blue, violet or even black to absorb the maximum amount of solar energy.

Eden's atmospheric pressure is fairly easy to judge. We've already seen that we can breathe more easily when the air pressure is around fifty percent lower than it is at sea level on Earth. This is the same level that we can find in mountain resorts. But we know we didn't evolve in the mountains. So we must have evolved on a planet where the air pressure at is around fifty percent lower than it is on Earth.

The lower air pressure and gravity, and the smaller height of the atmosphere (as measured from the surface of the planet to the edge of space) also means there wouldn't be as much air pressing down on us. So we would feel far more comfortable.

Eden must have an ozone layer that protects its surface from solar and cosmic radiation. But it's almost certainly thinner than the Earth's ozone layer. As Eden is as far from its star as Mars is from the Sun, it would need to allow a reasonable amount of UVB radiation to pass through it so we can create vitamin D. A thinner ozone layer would allow for this – and it also fits with the atmospheric pressure and oxygen level being lower than they are on Earth.

Climate

Eden's climate will be comfortable and temperate, even in its equatorial regions. Although it probably receives less solar energy than the Earth does, its denser atmosphere will keep it insulated. So its surface wouldn't be frozen, except perhaps at the poles.

As we've already noted, Eden probably doesn't tilt on its axis, so there will be no seasonal variations, and the climate will be stable all year round.

These two factors mean that Eden will be less prone to damaging weather events such as hurricanes, cyclones, tropical storms, tornadoes, monsoons, and floods. In fact, these might *never* occur. That would explain why we have no means of sensing them.

Eden's average temperature is almost certainly lower than the Earth's. As a result, crop yields will probably be lower too. This will help keep our population size in check.
    Another consequence of the cooler average temperature is that there might be fewer pathogens, such as viruses.

We know that when we first arrived on Earth we all had dark skin. The Europeans only developed white skin about 8,000 years ago[7-3]. Everyone on Eden would therefore have had dark skin too. But it was probably a natural response to the thinner ozone layer, rather than anything to do with the level of solar radiation.

It's highly likely that we had a thick coat of body hair when we lived on Eden. This would have kept us warm in the cooler climate, and it would have protected us against the higher level of UVB radiation. As we saw earlier, we would still have been able to create vitamin D, as it would have formed naturally when UVB radiation reacted with the grease (lanolin) in our body hair.

Soil composition

Eden's soil will be rich in minerals – particularly calcium – so it will be quite chalky. Most of the crops will be rich in minerals too. We can't get enough of these minerals from our food on Earth, and many people suffer from calcium deficiencies and associated health issues as a result.

Eden's soil might be rich in nitrogen, just like its atmosphere. We'll look at this in more detail in the section on plant life below.

Eden's soil will also be free of toxins and heavy metals – unlike Mars. One reason for this could be that the planet is so old that the toxins have been flushed away. Alternatively, the toxins might have become buried deep below the surface, with the top-most layers of soil having being replaced by organic matter. There certainly wouldn't be anything toxic like perchlorate, as there is on Mars.

An important consequence of the mineral-rich soil and plant life, and the lack of toxins, is that we would enjoy good health. Pregnant women wouldn't suffer from morning sickness, as their bodies wouldn't need to flush out any toxins.

Water

There must be plenty of fresh water on Eden, and we would have been able to drink it safely. We can drink pure spring water on Earth if it's filtered through rocks and it's free from microbes and other contaminants. But we can't safely drink from puddles, rivers and lakes as other animals can.

It's likely that most of our drinking water on Eden was filtered through rocks too. But we don't know whether *all* of it was free from contaminants, or whether we had evolved a degree of immunity to some of them.

We certainly haven't evolved any immunity to the contaminants on Earth. Thousands of people in developing countries die each year from drinking contaminated* water. That would be *extremely* strange if we had evolved on Earth.

Countries in the developed world sterilize their public water supplies, so the people living there will never develop any immunity to the contaminants* it originally contained.

> *When I say contaminants, I mean harmful microorganisms, the toxins they release into the water, naturally occurring chemicals and minerals, and things like animal waste and other decomposing matter. I'm not talking about man-made contaminants, and industrial or chemical pollutants. That's a separate issue.

Plant life

The edible plants on Eden will not be ones we recognize on Earth. In fact, the opposite will be true; we should *instinctively recognize* the edible plants on Eden, whereas we have to *learn* which ones on Earth are safe to eat, which ones are poisonous, and which ones are safe but "yucky."

Even babies would know the difference on Eden. And they would happily eat – and enjoy – the ones that are safe and most nutritious.

Here on Earth, children can't instinctively recognize *any* of the edible plants. Nor do they like the taste of most of the vegetables.

We've noted that the edible plants on Eden wouldn't have been particularly nutritious. The planet wouldn't have received enough solar energy to produce nutritious, high-yielding crops. But those crops would have adequately fed our small population. And they would have contained all of the vitamins, minerals and other nutrients we needed.

As we saw earlier, we are naturally herbivores; our digestive systems aren't capable of processing meat properly. On Eden, we would not have eaten meat. There would have been no need to, because the crops would have been packed with proteins and other nutrients. But when we first arrived on Earth, there were no such plants. We could only get those proteins and other nutrients from eating meat.

Eden's protein-rich, mineral-rich crops probably wouldn't grow on Earth, because its atmosphere, soil composition and climate weren't right. But it might have been possible to grow them on Mars – if its environment hadn't collapsed. And this might have been one of the reasons why the aliens considered Mars as a potential new home for us.

If we had eaten those same crops on Mars (and nothing else) it would have kept our population in check, just as it did on Eden.

And, of course, we would have evolved alongside the plant life on Eden. So we would not have been allergic to its pollen or suffered from hay fever there.

> We would almost certainly have eaten a much narrower range of foods on Eden than we do on Earth. We would have been more like the other animal species in this respect. This would have helped our children recognize which plants were good to eat, as they wouldn't have had as many to choose from as they do on Earth.

We've selectively bred most of the Earth's edible plants so they're more like the ones our tastes are attuned to – and presumably more like the ones that grew naturally on Eden. They might not look the same as the ones on Eden, but they might have a more familiar taste than the ones that evolved naturally here.

As well as changing their taste, we've also increased their nutritional levels, yields, resistance to disease, and other characteristics. Not only can we feed more people, we can supply them with far more calories than they need. As a result, our population has expanded massively, both in number and in waist size.

We may have begun selectively breeding the crops on Eden too. As a result, our population might no longer have been held in check. Our numbers may have begun to expand significantly – perhaps for the first time in our history. However, Eden might not have been able to support so many people – they may have drained its other resources. And that may explain why the aliens decided to remove us from Eden and prepare us for a new life in this solar system.

As we would only have eaten plants on Eden, it's unlikely that we would have farmed cattle. This would have helped keep the carbon dioxide and methane levels low.

Animal life

The animals on Eden almost certainly cover a similar range of classes and habitats as the animals on Earth. We might not recognize them – and we certainly wouldn't have eaten any of them. But they probably ranged from microorganisms to large beasts, small and large reptiles (perhaps even dinosaurs), as well as insects, birds, fish and other water-dwelling vertebrates and invertebrates, and small and large mammals. There might have been other hominin species like ourselves. There might even have been more advanced hominins than ourselves.

> The more advanced hominins *might* have been the aliens that brought us to Earth.

Predators

We have no natural means of defending ourselves against predators, or of escaping from them. This suggests that there weren't any predators on Eden when we evolved and lived there.

In the next chapter we'll take a look at planet Mars from a biological point of view. Could there have been any life there in the past, and could there be any today?

8
Life On Mars

As yet, there has been no official confirmation that life ever existed on Mars. But that doesn't mean it isn't there. There's every reason to believe life has existed on Mars for billions of years and still exists there today. However, it will almost certainly be anaerobic, primitive, microscopic, and deep underground. I sincerely believe it's there waiting to be found. We just need to look for it in the right place.

NASA's two *Viking* spacecraft, which landed on Mars in 1976, both detected evidence of life[8-1]. NASA scientist Dr. Gilbert V. Levin, who investigated the results of the Labeled Release (life-detection) experiment on the landers, reported that both craft had detected biological material in the Martian soil. NASA, however, dismissed the results as "false positives" and said the experiment had detected something that "mimicked life" but was not actual life.

Since then, Dr. Levin has been vocal in his criticism of NASA, saying the results were clear and conclusive. He has also accused NASA of deliberately delaying further life-detection missions, and said it had failed in its objective to seek out life on Mars.

He makes a good point. Since the *Viking* missions, NASA has avoided sending its landers to any of the areas of Mars that are most likely to support life – the so-called "special regions." Officially, this was to avoid contaminating them with microbes from Earth[8-2]. But this is hard to swallow. As Dr. Levin rightly said, one of NASA's primary objectives is to search for signs of life on Mars.

> Both *Viking* landers were sterilized before launch to prevent any chance of contamination – as was every lander that has been sent to Mars since then.

In any case, there is undoubtedly plenty of terrestrial life on Mars already. The Earth and Mars (and other planets in the galaxy) have been "swapping spit" for billions of years, as asteroids and meteorites strike the surface of each planet and send rocks spinning into space[8-3]. Geologists have found more than 120 rocks on Earth that originated on Mars[8-4], and there are probably hundreds more that are yet to be discovered. Dozens of rocks from Earth are likely to have reached Mars too – and perhaps many more than that. Most, if not all of them, will have carried some form of life. If the organisms were concealed and protected in crevices and fissures in those rocks, they might have survived the journey and established a new home on Mars[8-5].

Life has existed on Earth for over four billion years, and the Earth and Mars have been exchanging material throughout that time. It's unfeasible that no life-forms could have made the journey.

> It's also important to bear in mind that there were many more asteroids and meteors drifting around the inner solar system in those early days, shortly after the planets formed. Impacts would have been much more common than they are today.

The first life-forms that appeared on Earth around 4.2 billion years ago were anaerobic (non-oxygen-breathing). They could have survived – and even flourished – on Mars.

This began to change around 2.4 billion years ago, at the start of the Great Oxygenation Event. Single-celled life-forms in the oceans evolved the ability to produce energy from water and sunlight, and released oxygen as a waste product – a process known as photosynthesis. The level of oxygen in the atmosphere gradually rose over the next 400 million years until it reached about two percent (ten percent of the current level). This level would have been toxic to the anaerobic life-forms, and many researchers believe it may have killed off virtually all life on Earth at that time.

When life re-established itself, it was nearly all aerobic (oxygen-breathing) which means it would *not* have been able to survive on Mars.

So there was a finite window of opportunity, from 4.2 billion years ago to sometime after 2.4 billion years ago, when anaerobic life from Earth could have traveled to Mars and become established there.

> The first organisms on Earth lived without oxygen. Their respiration was inefficient, and they remained primitive. They probably drew their energy from hydrogen or nitrogen, but they only had around five percent of the energy of today's oxygen-breathing organisms.

> Oxygen allowed the aerobic organisms to develop a much more efficient form of respiration. It not only gave them twenty times more energy, but enabled them to transport it around and between their cells more easily. This led to the evolution of multi-cellular organisms and, eventually, to life as we know it today.

We need to reduce the window of opportunity slightly though, because you may recall that around four billion years ago one or more large asteroids hit Mars. If there was any life on the planet at that time, it would have been wiped out, in what biologists call a "global sterilization event." Some researchers believe Mars may have been struck by as many as twenty large asteroids over a 300-million-year period. So the planet may have remained sterile and uninhabitable for hundreds of millions of years.

It's reasonable to assume that Mars became habitable again around 3.7 billion years ago. So the window of opportunity for anaerobic life from Earth to have "seeded" Mars extends from around 3.7 billion years ago to around 2.2 billion years ago – which would have been mid-way through the Great Oxygenation Event. That's a total of 1.5 billion years. Hundreds of life-bearing rocks from Earth could have traveled to Mars during that time – and probably a lot more than that.

> Some anaerobic bacteria from Earth could have made the journey to Mars *after* the window of opportunity closed. In fact, they might still be making the journey today. Anaerobic organisms can be found in the mud at the bottom of the

> ocean; in geothermal hot springs, such as those in Yellowstone Park in the USA; and deep beneath the Earth's surface. There are nowhere near as many anaerobic organisms as there were before the Great Oxygenation Event, and (fortunately for us) there aren't as many asteroid and meteorite strikes either. So there's only a slim chance that any anaerobic organisms from Earth could reach Mars today – but it *could* still happen.

As we saw earlier in the book, Mars lost its magnetic field around four billion years ago. This was around the same time that the first major asteroid strike wiped out any life that may have been living there. As a result, the surface became exposed to radiation and solar winds, and it began to lose its atmosphere and surface water. Even so, it remained habitable (to primitive anaerobic organisms) for the next two or three billion years or more.

However, as the atmosphere dwindled, the surface froze, and the surface water was lost to space, drained underground, or was absorbed by the porous basalt rocks. Any remaining organisms would have had to move deep underground to survive.

As we saw earlier, there was a massive tsunami in Mars's northern around 3.5 billion years ago[8-6]. This not only proves the ocean still existed, it also indicates that the water must have been liquid and above freezing. Mars's oceans – and perhaps its other bodies of water – should therefore have been perfectly habitable to any of the Earth's anaerobic microbes that landed there[8-7].

There's another way in which life could exist on Mars, and that's if it appeared (or evolved) there first, before it appeared on Earth.

Mars formed around 4.6 billion years ago, and it would have taken around 200 million years for its surface to cool down enough for it to be habitable. So life could have become established there around 4.4 billion years ago[8-8]. Primitive anaerobic organisms could have evolved and lived there for 400 million years before they were wiped out by the first asteroid strike. This is interesting, because it means the first life-forms on Earth could have come from Mars.

If an asteroid or meteorite struck Mars around 4.2 billion years ago and sent a life-bearing rock spinning toward Earth – just as the Earth became cool enough to be habitable – it could explain how life on Earth got started.

Once the 300-million-year asteroid bombardment ended, there was a 1.5-billion-year window of opportunity for Earth's anaerobic life-forms to have seeded Mars with life.

Those life-forms *could* have been the descendants of ones that came from Mars and seeded the Earth 4.2 billion years ago.

When they arrived back on Mars, they could have evolved over the next two billion years, adapting to the changing conditions, and moving underground when the surface became uninhabitable.

> Unlike Earth, Mars didn't have an oxygenation event. Or at least we haven't found any evidence that it did. Later in the chapter, we'll consider what might have happened if Mars *had* had an oxygenation event.

If we detected life on Mars today, it would be impossible to tell whether it originated there or whether it originated on Earth billions of years ago and traveled there on rocks. (Or whether it originated on Mars, seeded the Earth with life, and then re-seeded Mars after the asteroid bombardment ended.)

Similarly, it's impossible to say whether life on Earth originated here, on Mars, or somewhere else.

If Mars *had* seeded the Earth with life (or vice versa), the life-forms on both planets would share a common basis to their DNA. This would also be the case if either planet (or both of them) had been seeded with life from elsewhere.

It's also worth bearing in mind that Venus was also habitable (to anaerobic life-forms) in those early days. And rocks carrying living matter undoubtedly ended up there too.

To everyone's surprise, a study in 2020 detected traces of phosphine gas in Venus's atmosphere. Phosphine is one of the signatures of life, so this discovery got everyone excited. It suggested there might be something living in Venus's clouds.

However, a follow-up study by another group found there was hardly any phosphine. This suggested that the original findings may have been an error[8-9]. We'll have to wait and see if this comes to anything.

> There might once have been life on Venus, but in my opinion there's unlikely to be anything living there today. Unfortunately, the acid rain has almost certainly erased any signs of previous life.

8. Life On Mars

Spacecraft on Mars have detected small amounts of methane[8-10][8-11]. Again, the presence of methane is often regarded as a signature of life, because living organisms produce it as a waste product. Methane decays quickly, so the fact that we can detect it on Mars today indicates that *something* is actively producing it. That doesn't necessarily mean there's anything living on Mars, though. It might just be a by-product of cosmic radiation reacting with chemical elements in the rock.

As we saw earlier, experiments on board the *International Space Station* have confirmed that some terrestrial life-forms can survive in space. Cultures of *Deinococcus radiodurans* bacteria were exposed to open space for over three years, and were still alive when they were brought back inside the space station. They had even multiplied and evolved while they were in open space.

Could these organisms (or their ancestors) have traveled on rocks and reached Mars? Yes, of course. Could they have survived there? Possibly.

The bacteria on the surface of Mars would almost certainly have been killed by bursts of ionizing radiation from solar flares. But any that had established homes underground could still be living, multiplying, and evolving there today.

Other extremophile species that traveled to Mars would probably have become dormant once they were exposed to anaerobic conditions and the vacuum of space. They would have ceased respiring, digesting, reproducing or moving, and they would have fallen into a state of suspended animation. Although they were non-functioning, they might not have died. Some of them could regain all of their functions if they're ever returned to normal conditions. And that could happen if we terraform Mars.

> In July 2020, Japanese scientists drilled a core sample in sediment under the sea and recovered microbes that had been dormant for 100 million years[8-12]. The microbes become fully functional when they were restored to normal conditions.
>
> It's highly likely that we will have terraformed Mars within the next 100 million years. If we do, a significant number of microbes from Earth that are living there today (in their dormant state) could become reactivated. Many of them could be prehistoric. Some of them might be harmful to us.

Any higher forms of life that were living in rocks that traveled to Mars probably expired shortly after they left the Earth's atmosphere.

We have no idea how life came to exist on either the Earth or Mars in the first place. Did the right chemical elements somehow self-organize themselves into a living, respiring, self-reproducing form? Did a strand of RNA, consisting of thousands of base pairs, all exactly in the right place, somehow grow in just a few million years while the planet was still cooling down?

It sounds highly implausible. The chances of such a thing happening over the course of *billions* of years, let alone a few million years, is infinitesimally small. Yet mainstream biologists insist it's what happened. And not only did it happen in an unbelievably short space of time, it also seems to have happened at the first possible opportunity.

> Francis Crick, the Nobel Prize-winning co-discoverer of DNA, famously said that the chances of life forming from molecules randomly crashing into each other was about the same as a hurricane hitting a junkyard and assembling a jumbo jet from random pieces of debris.

But what are the alternatives? Did a deity place fully-formed life-forms on the Earth? That sounds just as implausible as inorganic chemicals self-organizing into living cells in just a few million years. Yet it's the basis for most of the world's religions.

A more plausible theory is that life traveled to Earth (and Mars and Venus) on rocks from another planet outside the solar system. We've found rocks on Earth that contain the *building blocks* of life, and those rocks didn't come from any of the nearby planets; they came from somewhere much deeper in space.

Another theory is that the Earth was deliberately seeded with life by beings from another planet.

And there are several more outlandish theories, including one that suggests our universe might be a computer simulation and we're just bits of electronic data rather than physical beings[8-13].

But all of these theories simply push the main question further back in time.

If the building blocks of life ended up inside a rock, how did they get there? Presumably the rock was ejected into space from another planet that was struck by an asteroid or meteorite. But how did the building blocks of life form on *that* planet, and did life ever evolve there?

If life on Earth was created by a deity, how did that deity come to exist?

If we're computer simulations, who built and programmed the computer, and how did *they* evolve?

We could keep pushing the question back further and further in time until we reached the beginning of the universe, and still be none the wiser.

> In fact, we could push the question back even further than that, to the universe that came *before* this one. Or the one that came before *that* one…

Some biologists have suggested that Mars might once have been a lush, green paradise covered in vegetation. It's a nice idea, but it's extremely unlikely and there's no evidence for it.

If it were true, something equivalent to the Great Oxygenation Event that happened on Earth 2.4 billion years ago would have happened on Mars too. The level of oxygen in the atmosphere would have risen significantly, probably poisoning the anaerobic life and causing a mass extinction. The rocks on Mars would hold a record of this rise in oxygen, but none of the spacecraft we've sent there have detected it. We've also found no record of it in the rocks from Mars that have landed on Earth.

That's not to say it isn't there, though. If such an event happened, it would have taken place well over two billion years ago. So we might just have to dig a lot deeper under the Martian surface to find it.

Biologists who support the vegetation theory often cite the fifty-million-year head-start Mars had on the Earth. They also cite the fact that since Mars is a smaller planet it would have cooled significantly faster, meaning that it would have become habitable to life sooner.

But we need to remember that any life on Mars would have been wiped out by the asteroid that struck it around four billion years ago. And it would have taken a significant amount of time to recover.

If life evolved on Mars again after the impact, it would have been primitive, and hundreds of millions of years *behind* life on Earth.

Of course, if Mars was re-seeded with life from Earth once it had recovered from the asteroid strikes, it could have skipped hundreds of millions of years of evolution and caught up to the same level in one leap. But by the time that happened, the planet would have been in a state of terminal decline. Its magnetosphere was gone, its atmosphere and surface water were slowly disappearing, its surface was cooling and freezing – but also scorched by radiation and sterilized by solar flares every two hundred years or so.

As the surface was pretty much uninhabitable, any life that existed *must* have been aquatic, with most of it confined to the deepest parts of the oceans or in bodies of water deep under the ground. When the oceans eventually dried up, only the organisms living underground could have survived. In my opinion, there's no chance whatsoever that any vegetation ever appeared on the land.

But what if something like the Great Oxygenation Event *had* happened on Mars?

Let's imagine that cyanobacteria evolved in Mars's oceans, just as they did on Earth. And let's imagine that they developed the ability to photosynthesize, and they began producing vast quantities of oxygen.

As on Earth, there would have been a mass extinction that wiped out most life-forms. The oxygen would also have displaced some of the carbon dioxide from the atmosphere, reducing its insulating properties and causing the surface to cool down and probably freeze.

If the oceans had frozen, which they almost certainly would have done, the cyanobacteria would have died too, and the level of oxygen would have fallen again.

The good news is that the carbon dioxide would have eventually been restored to its previous level. The surface would have warmed again, and the oceans would have become liquid once more.

The problem is that by the time that happened, there would have been no anaerobic organisms left to repopulate the planet. Any life-forms that arrived on rocks from Earth would have been aerobic, and they wouldn't have been able to survive. So Mars would have been lifeless and barren – just as it is today.

> The Great Oxygenation Event on Earth led to an ice age, during which the whole surface may have been covered by a thick ice sheet. But as the Earth is closer to the Sun, the oceans remained liquid beneath the ice, and the aerobic organisms were able to survive.

If life exists on Mars today, it will be in pockets of liquid water deep underground, where it's shielded from radiation, UV rays, and the extreme cold. But it's unlikely to have evolved very far. The most advanced life-forms will probably be microscopic single-celled organisms.

There would also be lower organisms, which the more advanced ones might feed on. These might get their energy from chemical elements that are released when radiation breaks down the Martian rocks and soil. (Alternatively, the radiation might break the water down into its constituent molecules of hydrogen and oxygen, and the microbes might use the hydrogen as fuel.) There might also be hot water vents deep underground, similar to the ones we can find on Earth. These might contain dissolved minerals that the organisms can use as an energy source.

> The microbes might even have evolved a way of turning the high levels of perchlorate on Mars into energy.

Around four billion years ago, the rocks on Mars soaked up a huge amount of hydrogen, and much of it is still there. It extends miles below the surface, and there's enough to keep the microbes fueled with energy for hundreds of millions of years.

The microbes could thrive below the thick layers of ice at the poles, as the ice would help keep the hydrogen trapped underground. In fact, there might be a much better chance of finding life below the poles than anywhere else on the planet, as there's far more chemical energy there.

But there certainly won't be any *visible* signs of life on Mars – and nor could there ever have been.

Timeline of (possible) life on Mars

Life on Earth began about 4.28 billion years ago – or possibly earlier. Life may have gotten started on Mars before that, because the planet formed slightly earlier and cooled faster. Let's assume it began 50 million years earlier – or 4.33 billion years ago.

Any life on Mars was wiped out by an asteroid strike around four billion years ago, while life continued uninterrupted on Earth. The asteroid and meteorite bombardment on Mars continued for the next 300 million years, preventing life from becoming re-established.

Rocks carrying life-forms from Earth may have re-seeded Mars about 3.7 years ago. Or life may have evolved again naturally once conditions became suitable. If life on Mars had once been 50 million years ahead of life on Earth, it was now around 600 million years behind it.

But life on Earth was still extremely primitive at this point. The first single-celled organisms wouldn't appear for another 390 million years.

Although Mars may have been habitable to Earth's anaerobic microbes, the planet was slowly dying and its environment was collapsing. Let's assume it remained habitable until one billion years ago. How far might life on Mars have evolved by that point, assuming it remained 600 million years behind life on Earth?

The answer is: not very far at all.

Life on Earth had proceeded apace, and by one billion years ago it had become aerobic. That meant it had twenty times more energy than its anaerobic ancestors. The organisms were still single-celled, but they now had internal organs (organelles), and had split into three separate lineages that would become the plants, fungi and animals.

The first multicellular organisms would appear about 100 million years later.

None of this happened on Mars, because all life-forms (if they existed) remained anaerobic. They lacked energy, and since they also lacked nuclei and organelles, they had no means of transporting that small amount of energy around their cells. They certainly didn't have the energy to become multicellular, so they would have remained single-celled. If life ever existed on Mars, and if it exists there today, this is as far as it got.

An alternative timeline

Although we can find evidence that life first appeared on Earth around 4.28 billion years ago[8-14], this date isn't universally accepted by mainstream scientists.

The first *undisputed* evidence of life on Earth dates from 3.5 billion years ago. Let's see how this changes the timeline.

In this case, the first life-forms wouldn't have appeared on Earth until one billion years after the planet formed – and around 740 million years after it became habitable.

> 740 million years seems a much more reasonable period of time for inorganic elements to self-organize into primitive life-forms. Perhaps, like the mainstream scientists, we should ignore the evidence that it *actually* happened just 110 million years after the planet became habitable.

Let's assume it always takes around 740 million years for life to evolve on a planet that's capable of supporting it. We can discount the first 600 million years of Mars's existence, because life wouldn't have gotten started before the asteroid bombardment wiped everything out and set it back to square one. This also means there's no possibility that life on Earth could have been seeded by Mars.

If we assume it took another 740 million years for the first life-forms to appear on Mars once the asteroid bombardment ended, that takes us to 2.96 billion years ago. By that point, life on Mars would have been 540 million years behind life on Earth. But life on both planets would still have been anaerobic.

As we saw earlier, there's a possibility that rocks from Earth could have re-seeded Mars, and life could have appeared there earlier. But, regardless of where we are in the timeline, we can't escape the fact that life on Earth eventually became aerobic and multicellular, whereas life on Mars would always remain anaerobic, low-energy, single-celled, and primitive. There's a very low chance that it could have evolved any further without oxygen.

Even if Mars *had* had a Great Oxygenation Event, and even its life-forms *had* become aerobic, the first multi-cellular organisms would not have appeared by the time its surface became inhospitable to life.

In my opinion, we should focus on the original timeline that says life appeared on Earth around 4.28 billion years ago. First, because that's what the evidence says, and second because it presents us with the highly intriguing possibility that life on Earth *could* have been seeded by Mars.

In the next chapter we'll look at the future of Mars, including human colonization and the process of terraforming it into a facsimile of our home planet Eden.

9
Making Mars Habitable

For obvious reasons, we can't live on Mars as it is at the moment. Or at least, we can't live there without pressurized spacesuits, breathing apparatus, radiation and UV shielding, supplies of food and water, and so on.

The surface of Mars is devoid of life today, but there may be some extremely hardy microbes deep underground that we've yet to discover[9-1]. But that doesn't mean it was always that way, or that it will always be that way in the future.

Terraforming Mars would not only give us a back-up planet in case anything happens to the Earth (which it will), it would also give us the chance to build a better home – a slightly smaller version of our original home planet Eden.

As we work to make Mars habitable, we should *not* try to turn it into a facsimile of the Earth, because the Earth doesn't really suit us. Instead, we should try to create an environment that's perfectly suited to our physiologies, and fine-tuned to our needs.

Once Mars becomes habitable and we occupy it, we should start to become more like the people we were when we lived on Eden. All of the *bad* evolutionary changes we've undergone since we arrived on Earth should begin to undo themselves. But, hopefully, we'll keep the *good* evolutionary changes. We'll look at these changes in more detail in Chapter 11.

Let's quickly review what Mars was like in the past.

Historic Mars

The formation of the solar system

4.603 billion years ago
The Sun, Mars

4.543 billion years ago
Earth

4.503 billion years ago
Mercury, Venus, Jupiter, Saturn, Uranus, Neptune

The current evidence suggests that Mars was the first planet in the solar system to form, with the Earth forming about 60 million years later and the other planets forming 40 million years after that. Geologically speaking, these timescales are so close together that we could say they all formed at roughly the same time.

> But biologically speaking, the time difference is *hugely* important, because it means Mars could have seeded the Earth with life.

As it was so much smaller than the Earth and Venus, Mars's surface cooled faster. Life could have become established there just 250 million years after it formed. This would have been around 150 million years before life first appeared on Earth.

The atmosphere

Historic Mars had a dense atmosphere with a pressure 1.5 times greater than the Earth's. It was composed almost entirely of carbon dioxide, and there was little or no oxygen – but this was also the case on Earth at that time.

While the Earth eventually developed a breathable atmosphere, that didn't happen on Mars. If there was ever any plant life on Mars, it never evolved the ability to photosynthesize. So, throughout its entire history, the level of oxygen in Mars's atmosphere has remained minuscule.

Although Mars had a magnetic field when it formed, it was short-lived and only lasted for 500 to 600 million years. It disappeared completely around four billion years ago[9-2].

The methane

The presence of methane gas is often seen as an indicator of life. There is a very small amount of it on Mars today – around 0.4 parts per billion – which has gotten many people excited. Methane dissipates within twelve years of forming, and it's completely destroyed by sunlight within 340 years, so there must be a current or fairly recent source of it.

Most of the methane – or perhaps all of it – seems to be coming from an ice sheet near Gale crater[9-3] that was probably once a lake[9-4][9-5].

There's a small possibility that living microbes are producing the methane. Alternatively, it may be coming from deceased microbes that are slowly decomposing. So, while there might not be any life on Mars today, it's a possible indicator that there may have been in the past.

But we have to tread cautiously here, because methane is a simple hydrocarbon compound, composed only of carbon and hydrogen, and it can be produced chemically. For example, olivine rock produces methane when it's heated under pressure in the presence of carbon dioxide and water[9-6]. There could be a large olivine deposit deep under the ice sheet near Gale crater, and that might be where the methane is coming from. There will certainly be more than enough pressure, water and carbon dioxide there, and there will be plenty of heat if it's deep enough under the surface.

But even if olivine proves to be the only source of methane, that doesn't mean there was *never* any life on Mars. The microbes that once lived there could have decomposed and released their methane millions of years ago.

> There are microbes on Earth today that use methane as a source of energy.

Regardless of what's producing the methane, the ice sheet (and whatever lies beneath it) is definitely worthy of further investigation. It should be considered a priority for future exploration and deep drilling.

Even if the methane is being produced by a chemical process, we could still benefit from it. If we can locate the source and increase its rate of production, we might be able to use it as a fuel when we build habitats there. We might also be able to use it during the various terraforming processes that could make Mars habitable.

The level of methane in the Martian atmosphere varies with the seasons, peaking in late summer and early fall. Sensors have also detected occasional huge spikes in the level. As we don't know what's producing the methane, we can't tell what's causing the level to vary so much[9-7].

My hypothesis is that the ice sheet thickens in winter, blocking the release of methane. When it becomes thinner during the summer, it allows more methane to be released. The spikes in the level could be caused by holes occasionally appearing in the ice sheet.

The water

Studies of the former coastlines on Mars show it once had a massive ocean in its northern hemisphere that was around 2.5 miles deep. There were also smaller oceans and seas in the southern hemisphere. There were great lakes all over the planet, as well as rivers – some as large as the largest rivers on Earth[9-8][9-9]. There were also freshwater springs and outflows that extended for thousands of miles, as well as huge floodplains. In short, there was a *lot* of water on Mars.

Once Mars's atmosphere lost its density and its insulating layer of carbon dioxide, those massive bodies of water would have frozen, becoming ice sheets and glaciers. And as the pressure dropped, they would have begun to evaporate.

There's no longer any water on the surface of Mars. Until quite recently, we thought there was no water on the planet at all. But data returned from ground-penetrating radar systems on orbiting spacecraft has confirmed that around one-quarter of the water that once existed is still there – but it's now underground.

Why did the scientists think *all* of the water had been lost? When Mars lost its magnetosphere and the atmosphere was stripped away by the solar wind, they assumed the water went with it. They theorized that the water had boiled and evaporated as the atmospheric pressure reduced to almost nothing. It would then have been carried up into the atmosphere, photodecomposed into its component gases (hydrogen and oxygen) and been swept away into space with the other atmospheric gases.

But we now know this didn't happen.

Undoubtedly, *some* of Mars's water must have been lost via this process, but much less than the scientists originally believed.

There are two primary reasons why Mars's may have retained its water:

- The surface temperature may have fallen sharply when the insulating gases in the atmosphere were lost. The water would have frozen solid. With no rainfall to moisten it, the soil dried up and turned to dust, which covered (and protected) the frozen water.

- At around the same time that Mars's magnetic field disappeared, the planet was struck by one or more massive asteroids. The crust fractured and molten magma welled up over the surface. If the oceans had frozen by this point, they probably melted again. They might even have turned to water vapor for a short time before raining back down onto the surface. The magma cooled quickly, forming a type of basalt rock that's twenty-five percent more porous than normal. It acted like a massive sponge, and soaked up most of the water.

Radar data shows there's even more water trapped between layers of impermeable rock beneath the surface. There could be billions of gallons in these layers, and it may have combined with sand and dust to form a slurry.

It's important to understand that the events we're discussing here were not quick processes. The asteroid strikes, for example, probably occurred tens of millions of years apart. While each strike would have caused massive damage and destruction, and it might have taken the planet a million years or more to recover from each one, it *would* have recovered in the end.

The absorption of the oceans by the porous basalt rock would have happened glacially slowly too. And it would have been almost exactly balanced by the decrease in atmospheric pressure. Two billion years after Mars lost its magnetic field and it was hit by the asteroids, the oceans, seas and lakes continued to exist, and huge rivers continued to flow across the surface. They gradually became smaller, of course; and they would have sometimes frozen or become glaciers. But they continued to maintain a strong flow, erode the landscape, and form canyons.

We've found evidence of multiple wet and dry periods, with sea and river levels sometimes rising as well as falling. Sometimes the rivers (or glaciers) ran faster and grew longer, and sometimes they ran slower and became shorter, depending on the temperature and the amount of water or ice available. Sometimes they dried up completely, only to begin flowing again during the next wet period. Sometimes they froze solid, were completely covered by ice sheets, or only flowed underground.

Mars's underground water network is extensive and deep, and it remains pressurized even today. Liquid water is sometimes forced up through cracks in the crust, and small streams and patches of ice can be seen occasionally on the surface[9-10]. As the air pressure is so low, they don't last long or flow very far, and the water evaporates quickly.

> Although the frozen water at the poles is reasonably pure, data returned from orbiting spacecraft suggests that most of the other water on Mars is extremely salty – including the water that lies deep underground. Many researchers believe it's too cold and too salty to support life[9-11][9-12].
>
> We don't know if it was *always* too salty. It might only have become that salty when the surface water evaporated. And in that case, it might be acting as a preservative, preventing life-forms that once lived on Mars from decomposing.

> If the water is deep enough underground, it should warmed and kept liquid by heat from the mantel and core. In fact, it should be warm enough for life to survive there.

If we terraform Mars, some of the frozen water on or near the surface should melt. We already have desalination plants on Earth, and we could construct them on Mars too. So even if the water is salty, we could produce fresh, clean water for drinking and irrigation.

Melting the ice at the poles will cause the oceans and lakes to begin refilling, and the rivers will start flowing again. But most of the water now lies deep underground or trapped between layers of rock[9-13]. It's unlikely that the bodies of water on the surface would return to their former levels unless we pumped the water to the surface and prevent it from draining away again. We might also need to bring in more water from somewhere else. We'll come back to this later.

The geological events

Sensing equipment has detected occasional small tremors or "marsquakes" on Mars[9-14]. They're so minor (below magnitude 3.0) that few people would notice them, and they wouldn't cause any damage or injuries. Most of the quakes are probably caused by water pressure acting on cracks in the rock deep underground. This is almost certainly driven by tidal forces from Mars's largest moon, Phobos. A few of the quakes may be caused by meteorite impacts.

> There are more than a million tremors like this on Earth each year.

The surface of Mars was once divided into two tectonic plates, and there may have been earthquakes along their boundaries. But the plates didn't move around as much as the ones on Earth do, and the earthquake activity didn't last long. Mars's outer mantel cooled and solidified within a few hundred million years of the planet forming, and the two plates locked into place to form a single solid crust.

Mars once had numerous large volcanoes, but as the crust thickened and the mantel cooled, magma was no longer able to reach the surface. So the eruptions ceased and the volcanoes became extinct.

Human habitation on Mars

It's unlikely that we'll establish a permanent human colony on Mars any time soon. But I firmly believe we'll send a small team of astronauts on a mission to visit it within the next two decades. They'll spend an extended period of time in space on the outward and return journeys, and will almost certainly break the current record of 437 days for a single mission.

Engineers have proposed several means of getting them to Mars. The traditional approach relies on the Earth and Mars being at their closest points in their respective orbits. This happens every twenty-six months. The outward and return journeys would each take around nine months, and the astronauts would spend roughly five hundred days on Mars, remaining there until the two planets aligned again. The mission would take just under three years in total.

A better approach would be to fly past Venus first, and use its gravity to slingshot the spacecraft to Mars[9-15]. This would reduce the traveling time and the amount of fuel needed, and allow for shorter stays on Mars – perhaps as little as a month. Another bonus is that the planets line up in the right configuration every nineteen months.

The first astronauts on Mars will live in their spacecraft, which will have been heavily shielded against radiation for the journey. They'll probably be able to enlarge it using extensible panels and inflatable modules during the outward and return journeys and while they're on the Martian surface. The panels would be retracted and the modules deflated during the launch and landing phases.

The astronauts will need to take everything they'll need for the entire mission with them. But that might not be as much as you'd think. For example, they should be able to grow most of their food in an onboard hydroponic system. This will have been set into operation months before the astronauts depart from Earth, so they'd be able to begin harvesting food immediately. They would also recycle most of their water and air. We already have the technology to do this, as it's used on current space missions and on submarines.

The first few missions will study, collect and return rock, soil and water samples; look for current and past signs of life; and study the geology of the surface, looking at the different types and layers of rock, ice, former water courses and coastlines, and so on.

Later missions will investigate suitable places to establish long-term bases. These will almost certainly be in lava tubes or large cave networks. Underground bases will be better protected against solar flares and radiation, and the astronauts will have more room to move around than in their spacecraft. They'll also be cheaper and easier to build – and much safer – than habitats on the surface. The astronauts will need to seal the caves, tubes and tunnels, using airlocks at the entrances and exits. They'll then fill them with pressurized, breathable air and live and work more or less normally inside them.

> Research carried out on Earth suggests that the radiation levels inside the lava tubes would be around eighty-two percent lower than on the surface[9-16]. Additional shielding would still be needed in case a solar flare hit Mars. But as long as the astronauts had enough warning that a flare was about to hit them, they would only need a small, heavily shielded "bolt-hole*," which should be fairly easy and cost-effective to install. But we will have undoubtedly developed cheaper and more effective shielding materials by the time we're ready to build long-term bases on Mars.
>
> *An alternative would be to find much deeper caves or lava tubes, or excavate them ourselves. But as solar flares can cause huge spikes in radiation levels, they might need to be several miles deep.
>
> The only safe option for the longer term would be to restore or replace Mars's magnetic field and divert the radiation and energy bursts away. We'll look at some ways of doing this later.

One of the first unmanned missions to Mars will probably deliver one or more large vehicles to what will eventually become the base camp. These will enable the astronauts to drive around on the surface. The vehicles will probably be pressurized so the astronauts won't need to wear their spacesuits inside them. They might also have a new type of power pack that will enable them to run for years without needing to be refueled or recharged.

> By the time this happens, I'm sure we'll power vehicles and other electrical systems on Earth using similar technology. It might, for example, be a new type of nuclear power – much cheaper and more efficient than anything we have today, and completely safe.

Other unmanned missions around the same time will send the machinery, tools and supplies the astronauts will need to build their underground habitat. The astronauts will know before they set off from Earth that everything they'll need will be waiting for them, and that it's fully operational. They'll also have back-up supplies in case anything fails. And they'll have the tools, spare parts and training to be able to repair or adapt anything if it breaks down or needs reconfiguring.

Many science fiction writers have written stories about large cities on other planets, built beneath enormous glass, Perspex or Plexiglas domes. I don't foresee this happening on Mars within the next several thousand years – although it might happen after that. I don't believe the domes would be constructed from these traditional materials either. They will almost certainly be built from a newer material, such as silica aerogel, which we'll look at in a moment.

We will need to transport around one hundred million tons of material and equipment to Mars to build a city like this[9-17][9-18]. The massive dome that encloses the city would need to be constructed – which wouldn't be easy using our existing technology – and it would need to be filled with pressurized, breathable air. It would also need to shield the city against small meteorites, radiation, UV, and energy bursts from solar flares, with an extremely low probability of failure.

> A *zero* risk of failure would be preferable, but that might not be achievable in the case of the largest solar flares. While the possibility of failure remains greater than zero, there should be evacuation procedures, escape routes, and emergency shelters in place. Residents might have to move into underground bases or neighboring cities until their city or dome can be repaired or rebuilt.

How would we construct a dome large enough to enclose an entire city? How would it shield the city against radiation? While it would be much nicer to live in a domed city on the surface than in a cave underground, the cost would be astronomical, and the technology is well beyond us at the moment. For the next several thousand years at least, caves and lava tubes are the more practical options – and probably the only possible options.

> The massive cost of building a city on a planet hundreds of millions of miles away is another issue. Technology entrepreneur Elon Musk estimates it would cost around $140,000 per ton to transport the material from Earth. Sending the one hundred million tons of material required to build an entire city would cost $10 trillion to $14 trillion[9-19][9-20][9-21][9-22][9-23].

> It would be more realistic to downsize the city so it needed only one million tons of material to construct it. It would cost around $140 billion to send that amount of cargo to Mars.
>
> Once the first city was built, we should then be able to mine materials on Mars and build more equipment locally, rather than sending it from Earth[9-24]. This will allow us to build larger cities much more cheaply.

> Some NASA scientists are looking at the idea of *growing* the building materials on Mars. This would be easier than mining it, and much cheaper than transporting it from Earth. They suggest growing certain species of fungi, which could then be heat treated and compressed into building blocks[9-25].
>
> Another option would be to use chitin – a natural biopolymer that's found in the cell walls of fungi, the exoskeletons of insects and other arthropods, and fish scales.
>
> Researchers investigating the potential uses of chitin have found that when it's processed and combined with minerals in the Martian soil, they can create a building material that "feels like concrete, but much lighter."[9-26]

Let's jump forward a few thousand years or so. The domes and cities on the surface will probably be built by robots and intelligent machines[9-27]. The domes will be made from smart materials, be self-repairing, and able to withstand any seismic activity (such as marsquakes) and weather-related events (such as tornadoes) should they occur[9-28].

> The smart materials will be lined with sensors that monitor and detect any unusual activity. Should any damage occur, such as a breach in the dome, robots will be immediately assigned to repair it.

Jumping 100,000 years further ahead, we might not need the domes any more. The ultimate goal will be to fully terraform Mars, turning the surface from a dry, frozen, airless, lifeless desert into a place where we can live and thrive.

It will have a breathable atmosphere, a temperate climate, and oceans, lakes and rivers. The solar and cosmic radiation levels will be no higher than they are on Earth – and perhaps significantly lower. The citizens of Mars will be able to move around freely outside, just as we do on Earth, and grow a wide range of crops.

The one major difference between living on Mars and living on Earth is that we'd weigh significantly less on Mars because of the lower level of gravity. But, as we saw earlier, that should be a good thing.

Eventually, once we've constructed several small cities, and the total population exceeds one million, Mars should become fully self-sufficient and no longer need to rely on supplies from Earth.

Within two hundred years of that happening, I believe the two planets will become completely independent – politically, financially, and even, eventually, biologically.

As the citizens of Mars and their descendants adapt to the lower gravity, they'll find it increasingly uncomfortable to visit Earth. The gravity and air pressure will be too high for them, and they'll feel heavy and cumbersome. Their bodies, limbs and heads will hurt.

The strongest and fittest of them might be able to visit for short periods, but they won't want to stay for long. Eventually, as they adapt further to the lower gravity on Mars, it might become too dangerous for most of the population to visit the Earth at all – or at least not without wearing negative pressure suits[9-29] and other protective equipment.

On the other hand, people from Earth *would* be able to visit to Mars, or even settle there permanently. But again, once they'd become acclimatized, returning to Earth would be a painful experience – and potentially a deadly one – depending on how long they'd spent on Mars.

Ultimately, as they evolved with different levels of gravity and other environmental conditions, the citizens of Earth and Mars would become two separate sub-species and look different from each other[9-30][9-31].

Later in the book, we'll look at the changes our bodies will undergo as we adapt to living on Mars.

Terraforming Mars[9-32]

Fully terraforming Mars will take tens of thousands of years at the very least. More likely, it will take as long as 100,000 years. Some researchers doubt it's even possible[9-33].

I believe we *will* eventually develop the technology, materials and equipment we'll need to do the job. But I don't believe we'll even begin the terraforming process for several thousand years. It's not just about developing and testing the theories, and creating the materials, equipment and techniques to turn those theories into reality, it's about the astronomical cost of doing it, as well as the political will, and the environmental will. Many people will argue that we shouldn't be meddling with another planet at all when we can't even look after this one properly.

> But if we can develop the technologies to fully terraform another planet, we should be able to use the same technologies to repair the Earth.

> Many people argue that we should not consider terraforming Mars until we've established once and for all whether it ever hosted life – or whether it still does. If life survives there today, then it's an important ecosystem that we should protect. If it hosted life in the past, then it's an important historical site that we should protect. We should only start the terraforming process once we've confirmed that it's devoid of all life-forms. (At least until we have no other choice and the Earth can no longer support us.)

There's also the argument that we shouldn't terraform Mars into a back-up planet for the Earth, because we would then neglect the Earth.

Campaigners argue that we should focus our energies on repairing the problems with the Earth first, and *then* think about colonizing Mars and other planets, rather than making the colonization of Mars our priority[9-34].

While I agree with this to some extent, I believe we should be doing *both* of these things.

From an ecological point of view, I'm proud to have devoted the last two decades of my life to helping to clean up the world's oceans – and I continue to do this for several months each year. But I also know our time on this planet is limited.

The Earth *will* become uninhabitable – not just to us but to every other living species – in around 600 to 800 million years' time (or less). It's vitally important that we push ahead with our plans to colonize other planets, starting with Mars. It's our only chance of surviving as a species.

Having said that, we will of course have evolved into a *different* species by then. But that's a separate issue too!

Let's consider what we would need to do to terraform Mars into a planet we can live on.

Reinstating the magnetosphere[9-35]

As we've already seen, our first priority must be to reinstate Mars's magnetosphere. This is vitally important, because any attempt to reinstate the atmosphere would fail if it was continually eroded by the solar wind. The Martian atmosphere is wispy thin, and until it reaches a critical mass, it will keep getting stripped away.

Reinstating (or replacing) the magnetosphere would also increase the level of protection from solar and cosmic radiation.

Researchers have proposed two potential solutions. The most promising of these – and the one I support – would be to place a powerful magnetic dipole shield in space between the Sun and Mars. Of course, we don't have the technology to build one yet, or to power it – and it would need a *huge* amount of power.

> The shield would need to be placed at Mars's L1 Lagrangian Point[9-36]. This lies 672,517 miles (1,082,311 km) from Mars, and is the point where the gravitational pull of Mars exactly balances the gravitational pull of the Sun. This would ensure that the magnetic shield always remained in the right place as it orbited Mars.

> We should be able to increase the shield's power whenever a solar flare threatens Mars. This should prevent even the largest flares from damaging electronic systems – and life-forms – on the planet. If the system works, we might be able to use a similar system to protect the Earth, temporarily boosting *its* natural magnetic field as and when it's needed.

Another option would be to build several superconducting rings around Mars's circumference. These would be located on the planet's surface. They would need to be refrigerated and electrically powered, but they *could* (in theory) be built using our existing technology. They would of course take an enormous amount of manpower (and machine power) to build on Mars. And the cost would be astronomical. It might take us hundreds of years to build each ring, and we would need to house the construction workers on Mars while they built them. And, like the magnetic dipole shield, the rings would need a *huge* amount of power to keep them running.

While the superconducting rings would be easier to build and maintain using our current technology, I can't see them being built for thousands of years. We will almost certainly have developed the technology to build and power the magnetic dipole shield by then, and I believe it would be a better option.

Once we've developed the technology, the shield – and at least two or three back-up units – would only take a few years to build. The shields could also be built on Earth – or in Earth orbit depending on their size and our ability to ferry large objects into space.

We wouldn't be able to activate and test the shields until they were well out of range of the Earth. A good first test might be to see if we can divert the solar wind, radiation, and energy from solar flares around our own Moon, and make *its* surface a safe place to live[9-37][9-38]. If we found any problems with the shields, we could easily bring them back to Earth to fix them.

The Moon once had a magnetic field that was at least as strong as the Earth's is now. It lasted from around 4.2 billion years ago to around 3.4 billion years ago. At that time, the Moon was twice as close to the Earth as it is today. At the same time, the Sun was around 100 times more active than it is today, and the streams of particles it emitted could have stripped the Earth's atmosphere away. Researchers believe the Moon's magnetic field may have combined with the Earth's to give it the extra protection it needed. They suggest that as we search for habitable planets, we should pay particular attention to those with large moons[9-39].

We would need to be able to activate a back-up shields within moments of one failing, to prevent any harm coming to the citizens (and electronics) on Mars. The fault detection and switchover would need to happen automatically, as it could take almost an hour to detect the failure on Earth and send a signal to activate a back-up device.

> Once a shield had been activated, we probably wouldn't be able to get anywhere near it, as it would emit a massive electromagnetic field. If a shield needed maintenance, it would need to deactivate its magnet, activate one of the back-up units, alert its controllers, and relocate itself to either Earth orbit or Mars orbit where engineers or robots could work on it safely. This would need to be a failsafe process that happened automatically, even if every other system on board the shield failed.
>
> But in the event of *total* system failure, we might need to send an unmanned craft with no magnetic parts to retrieve the faulty shield.

Reinstating the atmosphere

Some researchers believe that once Mars's magnetic field has been restored and the solar wind has been diverted away from the planet, the atmosphere should begin to reinstate itself without any further intervention.

Frozen carbon dioxide in the ice caps should begin to sublimate (change from a solid into a gas) and be released into the atmosphere. The thickening atmosphere would then trap increasing amounts of solar radiation and warm the planet's surface, causing even more carbon dioxide to be released. The ice at the poles would also begin to melt, partially reinstating the oceans[9-40].

Researchers have run simulations which show that within just a few years, the atmospheric pressure would be about half what it currently is on Earth. This would be perfect for us – it's the same pressure as our original home planet Eden, and, as we saw earlier, it's the same pressure that we can find in habitable mountainous regions on Earth.

But we still wouldn't be able to breathe the air on Mars. The high level of carbon dioxide would be deadly, and there wouldn't be any oxygen. We'll address these issues below.

Initially, the goal might be to raise the atmospheric pressure to 2.8 psi (pounds per square inch). This is the level where we could survive without needing to wear pressure suits. But we would still need to wear tight-fitting masks to deliver oxygen and keep the carbon dioxide out.

The next goal would be to raise the air pressure to around 10 psi – similar to that found in habitable mountainous regions on Earth, and at sea level on Eden. In my opinion, we should not make the air pressure any higher than this.

Several well-known people, including the space and technology entrepreneur Elon Musk, have suggested that the best way to begin the terraforming process would be to heat Mars up. This would force trapped carbon dioxide to be released from the ice and rock[9-41]. Musk has proposed detonating nuclear bombs at the poles to achieve this. Other theorists have suggested placing large mirrors in orbit around Mars to focus solar radiation onto the surface. The mirrors would of course be the more sustainable and less damaging option, so that's the option I support.

> Elon Musk's idea is to detonate a continuous stream of low-fallout nuclear bombs[9-42] above Mars's north and south poles, effectively creating two artificial suns[9-43].

> Mathematician Robert Walker has calculated that we would need to detonate 1,728 nuclear bombs at each pole every day to raise the temperature on Mars to a habitable level. That's one bomb every fifty seconds for about seven weeks.
>
> As there are two poles, we would need to detonate 3,456 bombs each day for seven weeks – and about 169,000 bombs in total[9-44][9-45].
>
> By way of comparison, the USA currently has around 3,800 nuclear warheads in its entire arsenal. That would be enough for the first day.
>
> Once we'd heated Mars to the required temperature and released enough carbon dioxide to reinstate the atmosphere, we could slow down the rate of the explosions. But it's unlikely that we could ever halt them completely.

> If we placed a mirror in orbit around Mars to reflect solar energy onto the surface, it would need to be about 150 miles (240 km) in diameter. In reality, of course, we would use hundreds of smaller mirrors, each around 500 feet (150 meters) in diameter[9-46][9-47].

Another approach would be to cause climate change on Mars in the same way that we've caused it on Earth. For example, we could burn fuel in hundreds of power stations and release the waste carbon dioxide and other gases into the atmosphere, where they would trap the Sun's heat. As well as helping to reinstate the atmosphere, this would give us a planet-wide electricity network. Among other things, we could use this to power the superconducting rings that would restore Mars's magnetic field.

> One researcher estimated we would need 245 power stations, each generating around half a gigawatt of electricity.

While several researchers have suggested this approach, I'm not sure where the fuel would come from. None of the conventional fuels we use in power stations on Earth would be available on Mars. There are no reserves of coal, oil, natural gas, wood, or biomass. It's far more likely that we would generate power there via solar, geothermal or nuclear means, but none of these creates much carbon dioxide.

We could possibly extract carbon from the carbon dioxide that's already on Mars and use that as a fuel. But as we're trying to *create* carbon dioxide at this stage rather than deplete it, that kind of defeats the purpose.

> Converting the carbon dioxide into carbon and oxygen will be useful in the *next* stage of terraforming, when we detoxify the atmosphere.

One other solution would be to reduce Mars's albedo – the amount of light its surface reflects. One way of doing this might be to collect dark dust from Mars's two moons and spread it over the surface. The surface should then absorb more heat from the Sun and warm up.

While this is a great idea in theory, it would be highly impractical to transport enough material from the moons. And, of course, it's another thing that's way beyond our current level of technology. Another issue is that the frequent global dust storms on Mars would blow the dark dust around and quickly bury it, so it would need to be replenished.

Even if this idea worked for a short time, the dust storms would probably become more frequent as the surface heated up, and that would just make the problem worse. We would need to keep bringing in more and more moon dust, in the hope that the surface would eventually heat up enough to melt the ice and restart the water cycle. The dust would then become moist, clump together, and stop blowing around during the storms.

The big problem with all of these amazing ideas is that they were proposed as *primary solutions*. That means they would be carried out as the first step in the terraforming process, ahead of anything else.

But as we've already seen, if Mars's magnetosphere isn't reinstated or replaced first, the carbon dioxide that's released as the planet heats up will be lost to space. So it too would need to be constantly replenished[9-48]. The primary solution must be to address Mars's lack of a magnetosphere first.

> Detonating more nuclear bombs on Mars once it had been terraformed wouldn't be the best idea, especially if it was already inhabited. Again, the orbiting mirrors would be the best option … until the carbon dioxide runs out.

9. Making Mars Habitable

This leads us to another serious problem: even if we can somehow release *all* of the carbon dioxide that's currently trapped in the rocks and ice on Mars, there might not be enough of it to terraform Mars.

Some geologists estimate there's only enough carbon dioxide to raise the atmospheric pressure to about seven percent of the Earth's pressure. That would be less than 1 psi. So we would still need to wear pressure suits to survive.

We would need to raise the pressure to at least forty to fifty percent of the Earth's pressure (around 10 psi) to live comfortably and be able to breathe without assistance.

An obvious solution would be to supplement the carbon dioxide with another gas. But in fact, there's a much better way, and that's to not use carbon dioxide at all. After all, it's toxic to humans, and we would have to remove most of it later in the terraforming process so we could breathe.

Some researchers have suggested it may be possible to terraform Mars in just fifty years using halocarbon gas. This is a powerful greenhouse gas made by combining carbon with a member of the halogen group of elements (fluorine, chlorine, bromine, iodine and astatine). They suggest that an industrial plant could be constructed on Mars to create the halocarbon gas and pump it into the atmosphere. It would need about a gigawatt of electricity to power it – which could be easily supplied by a moderately large nuclear power station using today's technology.

> To create the optimum mix, and terraform Mars even faster, we could combine the halocarbon with sulfur. It would make the atmosphere *deeply* unpleasant for a while, but we could fix that later.

But once again, we would need to remove the halocarbon from the atmosphere later in the terraforming process to make the air breathable. This could be difficult. Many forms of halocarbon are toxic, it doesn't break down naturally, and its byproducts are poisonous and corrosive.

> Some species of desulfitobacterium might be able to break down the halocarbon. Researchers are investigating their potential. We could spray Mars with the bacteria[9-49], leave them to get on with their work, and come back when they'd finished. The problem is that while they might remove the halocarbon, they'd also create millions of tons of carbon dioxide. So we'd then have to find a way of removing *that*. Again, bacteria could provide the answer, but they would need to be man-made ones rather than anything that occurs in nature[9-50].

> This also provides us with a possible solution to the problem we just looked at, involving the shortage of carbon dioxide on Mars. We might be able to use bacteria to create it.

Another issue is that any gas we use to replace the carbon dioxide or halocarbon would not only need to be non-toxic and breathable, it would also need some insulating, heat-retaining properties, otherwise Mars would cool down again as soon as we removed it.

A better solution might be to use perfluorocarbon (PFC). It would achieve the same effect as carbon dioxide or halocarbon, but unlike its more harmful relative chlorofluorocarbon (CFC), it's non-toxic, long-lasting, and doesn't deplete the ozone layer[9-51]. However, Margarita Marinova, the MIT student (and now NASA engineer) who proposed this solution, said she would only use it to heat up Mars until it began releasing its own carbon dioxide. After that, she would only use PFC to "plug any gaps."

As we've already seen, there are some major problems with this approach. The first is that the "gaps" she talks about could be massive – there might not be enough carbon dioxide on Mars. The second problem is that even if there *is* enough carbon dioxide, or if we supplemented it with PFC, we'd still be left with a toxic, carbon dioxide-rich atmosphere.

Another problem we haven't considered yet is that we'd need to build a significant number of industrial plants on Mars to pump enough PFC into the atmosphere. This is beyond our current ability. But it might not be beyond the realms of possibility in the future, as we could send robots to construct them.

> We would also need to cover Mars with halocarbon factories if we used that option instead.

Researchers have suggested that if we used PFC, it would take around eight hundred years to raise the temperature of Mars to the point where the ice would begin to melt. Marinova says they've overlooked the fact that Mars's own gases (particularly carbon dioxide) would be released as the temperature rose. The atmosphere would therefore retain more solar energy, and it would take significantly less than eight hundred years to reach the required temperature.

In fact, if we combined PFCs with halocarbons, raising the surface temperature and creating a sufficiently dense atmosphere might only take a century or so.

Detoxifying the atmosphere

Let's jump even further ahead and assume that Mars's atmosphere has been fully reinstated. The atmospheric pressure is now around fifty percent of the Earth's – which is the perfect level for us. The problem is that it's composed almost entirely of carbon dioxide. We'll need to get rid of most of the carbon dioxide, while retaining its insulating properties and maintaining the atmospheric pressure.

This could be extremely tricky for three reasons:

- First, we would need to extract the carbon dioxide from the atmosphere and convert it into something else. The best option might be to break it down into its separate components: carbon and oxygen. The carbon could be stored safely and used as a building material or fuel. And we'd need plenty of oxygen anyway, as it only exists in trace amounts currently.

- The second problem is that splitting carbon dioxide into carbon and oxygen takes an enormous amount of energy – and we'd need to split billions and billions of tons of it. Fortunately, by the time we're ready to begin this process, we should have developed a large enough power source to drive the magnetic dipole shield or superconducting rings that would replace or restore Mars's magnetic field. We should be able to use the same power source to split the carbon dioxide.

    We won't be able to begin the terraforming process *at all* until we've developed this power source.

One solution to both of these problems would be to take the natural approach. We could use vegetation to process the carbon dioxide into oxygen, just as it did on Earth over two billion years ago. Robots could plant and maintain the vegetation. The main downside is that this process would take at least 100,000 years to work[9-52].

- The third problem, which we've already discussed, is that as we remove the carbon dioxide from the atmosphere, its insulating properties will be lost. The surface will cool down again, and eventually freeze over.

We'll obviously need to replace the carbon dioxide with something else as we remove it, to maintain the atmospheric pressure and retain its insulating properties. It will need to be a gas that we can breathe safely, but which is, ideally, also a greenhouse gas with similar insulating properties to carbon dioxide. The only gas that fulfills both of these requirements is water vapor[9-53].

Of course, if we used nothing but water vapor, we would literally be walking around with our heads in the clouds. The clouds would be thick, wet, and at ground level. It would be horrible to breathe – we'd feel as if we were drowning – and it would be so foggy that we wouldn't be able to see our hands in front of our faces.

So we would need to limit the amount of water vapor in the atmosphere, keep it well above ground level, and replace a significant amount of it with something else. I would suggest using nitrogen – the same inert gas that makes up most of the Earth's atmosphere.

The main problem with this approach is that we'd need a lot of nitrogen, and there's hardly on Mars. There's less than three percent in the atmosphere today, and as we increase the amount of carbon dioxide and water vapor during the terraforming process, that proportion will reduce even further. The terraformed Martian atmosphere will need a similar level of nitrogen as we have on Earth today: around seventy-eight percent. But where would we get it from?

We couldn't ship it over from the Earth, as we'd need billions of tons of it. It would be impractical to transport that amount, even if we compressed it into liquid form. It would also deplete the Earth's atmosphere. This might actually be the hardest part of the terraforming problem to solve.

In fact, we're not even sure how the Earth got its nitrogen[9-54]. There are theories that it might have come from the Sun or from comets. But the nitrogen in the Sun has lighter isotopes than the nitrogen on Earth, while the nitrogen in comets has heavier isotopes. Earth's nitrogen *might* be a mixture of both types, but we don't know for sure.

Does it matter if the nitrogen we use to terraform Mars is heavier than the nitrogen on Earth? Probably not[9-55]. DNA seems to work just the same in both types. It changes slightly though; it leaves clear markers so you can tell which type of nitrogen the organism lived in.

The best solution might be to find a comet or meteorite that's rich in nitrogen, divert it, and crash it into Mars[9-56].

Of course, that's yet another thing that's way beyond our current level of technology. So we won't be able to begin terraforming Mars any time soon – and probably not for thousands of years.

An alternative approach would be to crash an *ammonia*-rich comet or meteorite into Mars. Ammonia is another effective greenhouse gas that would kick-start the terraforming process. Once it had done its job, we could use an industrial process to extract it from the air and pump it over heated metal ions to release nitrogen and hydrogen.

> If we waited long enough, the ammonia would naturally photodecompose into nitrogen and hydrogen. But it would be a *really* long wait, because ammonia is a highly stable compound.

If we were going to use an ammonia-rich comet or meteorite rather than a nitrogen-rich one, we'd be better off converting the ammonia into nitrogen right from the outset, and using water vapor as the greenhouse gas. That way, we'd get to the breathable atmosphere stage much sooner.

My solution to the problem of how to detoxify the Martian atmosphere would therefore be to:

- develop an ultra-efficient energy source to power industrial processes and electronic systems, and the systems that would replace Mars's magnetic field

- locate and divert a nitrogen- or ammonia-rich comet or meteorite and crash it into Mars

- use the new energy source to remove the carbon dioxide from the atmosphere by splitting it into carbon and oxygen, and at the same time split the ammonia into nitrogen and hydrogen

- plant crops to help take up more of the carbon dioxide and produce oxygen

- cover the sky with a thick blanket of clouds (water vapor) to trap solar energy, increase the surface temperature, and minimize heat loss

Maintaining the temperature

Increasing Mars's surface temperature is one thing, but maintaining it is another. Mars receives less than half the amount of sunlight that the Earth does. If we gave it the same atmosphere as the Earth, the average surface temperature would only be about -60°F (-50°C) – about the same as Antarctica. There would be no liquid water, and no plants or animals, as it would be too cold.

> We're planning to give Mars an atmosphere with only half the pressure of the Earth's. So the problem would be even worse.

> The Earth's average surface temperature is around +61°F (+16°C).

The best solution would be to boost the meager sunlight using some of the orbiting mirrors we used earlier in the terraforming process.

Restoring the liquid water

With the terraforming process well underway, and Mars's surface temperature on the rise, the ice at the poles would begin to melt. There's enough water at the poles to fill the oceans to around one-seventh of their former levels. So they would be significantly shallower than they were during Mars's prehistoric era.

The water that lies deep below Mars's surface might remain frozen for millions of years, and some of it might never melt at all. Even if it melted, most of it would remain underground in the porous basalt rock or trapped between impermeable layers of rock.

If we wanted to restore the oceans and seas to their original levels, refill the lakes, irrigate the land, and make the rivers flow again, we would either need to pump the water to the surface (and prevent it from draining away again) or bring in more water from another source.

Once again, crashing a comet or meteorite into Mars might be the best solution. If the same nitrogen-rich or ammonia-rich comet or meteorite we considered using earlier was also packed with water ice (or hydrogen)[9-57], we could solve two problems at once.

Other possible (though not currently practical) options include shipping water ice from one of the moons of Jupiter or Saturn; or capturing a small, icy moon from the outer solar system and crashing it into Mars.

Insulating the atmosphere

As well as restoring the liquid water on Mars, we would also need to find a way of pushing water vapor into the atmosphere and keeping it there, so that it would act as an insulating layer. This happens naturally on Earth, where clouds form as part of the water cycle. We should be able to replicate this cycle on Mars.

However, Mars would need proportionately more water vapor in its atmosphere than the Earth does, because it receives less solar radiation and it's more prone to heat loss.

In cold, dry areas on Earth, the proportion of water vapor in the atmosphere can be less than one percent, while in humid, tropical areas it can be as high as four percent. The terraformed Mars will need a minimum of four percent over the whole planet. That means its atmosphere will need to hold at least 10 million billion gallons of water.

> That might sound like a lot, but there's around 37.5 million billion gallons of water in the Earth's atmosphere.

Restoring the ozone layer

Mars has three ozone layers, but they're around three hundred times thinner than the Earth's single layer. They are far too thin to filter out ultraviolet radiation from the Sun, which passes straight through to the surface and destroys organic molecules. Most researchers believe nothing can live on the surface because of this[9-58].

Once we've restored Mars's liquid oceans and given it a reasonably dense atmosphere with a high proportion of oxygen, the ozone layers should begin to thicken all by themselves. They'll be extremely fragile for hundreds of years though, and it will be vitally important that we do nothing during this stage of the terraforming process that might disrupt or destroy them. We *definitely* shouldn't use compounds like CFCs, which caused severe damage to the Earth's ozone layer.

> Worryingly, some ill-advised researchers *have* proposed using CFCs to cause a greenhouse effect on Mars. This could ruin the entire terraforming process, and it might *never* become habitable.

Detoxifying the soil

The soil on Mars contains a high level of perchlorate, which makes it too toxic for plants to grow. However, perchlorate dissolves readily in water, so it's reasonably easy to deal with: we just need to flush it out. This would be an intensive process, but it could be easily mechanized and automated, and we could process just a few acres at a time.

Dealing with the contaminated water created by this process would be fairly straightforward too. An industrial process such as ion exchange or reverse osmosis could turn it into brine. Or a biological process – using bacteria from Earth – could turn it into chloride and oxygen.

While it might take several centuries to decontaminate the whole of Mars, we should be able to remove the perchlorate from an area the size of a large farm in just a few weeks. We could then repeat the process on the next area of farmland, and keep going until we had decontaminated the entire planet.

> Two-thirds of Mars's surface would be covered in water by this point, so there wouldn't be as much land to decontaminate as there is now.

> The surface water could become contaminated with perchlorate, of course. We might need to reduce the level of chloride if anything is to survive in it. We could use the same systems that we use in desalination plants on Earth (ion exchange and reverse osmosis). But it might take thousands of years (or more) to process all of the water on Mars.

Growing crops

There's a great deal of debate about whether enough solar energy reaches Mars for crops to grow[9-59]. The general consensus seems to be that a *few* crops would grow, in a limited way, but they would produce low yields, might not ripen, and would be low in nutrients.

> The same might be true of our home planet, Eden. As we saw earlier, the evidence suggests our population was small because of a lack of nutrition.

We should be able to modify certain crops to produce higher yields in low-light conditions. But a better solution would be to grow them under artificial lighting. We already grow many types of crop commercially on Earth under high-energy LED lighting. We could also use this tried-and-tested system on Mars.

> The new, ultra-efficient (and cheap) energy systems we develop to power Mars's magnetic shield and the other terraforming processes could also light its farms and food production facilities.

Some people have expressed concern that the Martian soil might be too acidic for crops to grow. In fact, the opposite is true: the soil is slightly *alkaline* – its pH ranges from 8.0 to 9.0 and averages around 8.3.

Fortunately, it has just the right pH and mineral content to grow many crops from Earth. Asparagus, beans, chives, corn, cress, leeks, peanuts, peas, quinoa, radishes, rocket, rye, soybeans, spinach, sweet potatoes, tomatoes, turnips and wheat would all grow in this type of soil[9-60][9-61] – once the perchlorate was removed.

In experiments that simulated the Martian soil (albeit at terrestrial levels of light), these crops produced about half the yield they would have produced on Earth. Over time, we should be able to modify the soil – and the crops – to increase the yield and match the levels we see on Earth.

We would need to analyze the Martian soil thoroughly before planting any crops. If any heavy metals were present, for example, they could be taken up by the plants, and they would become toxic. If that were the case, the soil might need further decontamination to make it safe for farming.

If we wanted to grow crops that need more acidic conditions, such as fruit trees, we would either need to treat the soil with chemicals, or grow the crops in containers. Increasing soil's acidity is a fairly straightforward process. It can be achieved by adding things like sulfur, aluminum sulfate, iron sulfate, or acidifying nitrogen. We might be able to mine some of these things on Mars, or create them as by-products of terraforming or other industrial processes. As a last resort, we could ship them from Earth. But this would be impractical on a large scale, and impossible on a global scale, because it would deplete the Earth's resources. A better option would be to use bacteria to increase the soil's acidity.

Many crops could be grown in large-scale hydroponic facilities, where they wouldn't need any soil at all. We already do this on Earth. The crops only need a substrate, such as coir or rockwool, and water laced with minerals and nutrients. Crops that grow well in these systems include cucumbers, herbs, lettuces and other leafy greens, peppers, and strawberries.

In fact, the first crops to grow on Mars will almost certainly be grown hydroponically, underground, under LED lighting.

> Astronauts on board the International Space Station have grown crops hydroponically – and eaten them. Astronauts traveling to Mars will use a similar system to provide food for their mission.

> Researchers say a self-sustaining hydroponic system large enough to feed one million people could be transported to Mars quite easily. And it could be done using our current technology.

Eradicating the global dust storms

The restoration of Mars's surface water and moist atmosphere should put an end to the dust storms, as the dust will become clump together into soil. Eventually, we might cover much of the planet's surface in grass and other vegetation to help lock the soil in place.

But the restoration of the water cycle will create an active weather system, and that could cause storms and even tornadoes. We've already seen small "dust devils" on Mars. However, as less solar radiation reaches Mars compared with the Earth, its weather system should also be less active. Damaging or life-threatening conditions are much less likely to occur.

We should even be able to control the weather to some extent. For example, if we used orbiting mirrors to reflect more sunlight onto Mars, we could turn them away from the surface during times of danger. This would remove the storm's source of power, and it should dissipate fairly quickly.

The first colonists

The first *permanent* inhabitants of Mars probably won't be human. They might not even be visible to the naked eye. Life existed on Earth for billions of years as simple microbes and single-celled organisms, and they did a great job of terraforming the planet for the more advanced life-forms that came later. It would make sense to duplicate that process on Mars by sending useful bacteria, viruses, and fungi that support many of life's processes[9-62].

Selecting the right ones will be a challenge though, and there will, no doubt, be considerable differences of option. As we've already noted, many objectors will argue that we shouldn't be sending them at all, as we would be contaminating another planet[9-63].

One of the biggest problems with using microorganisms to terraform Mars is that it might take them billions of years to do the job – just as it did on Earth. But, on the other hand, it might not. The first microorganisms on Earth were anaerobic and extremely primitive, so they were slow and inefficient – but they still got the job done in end. If we send aerobic microorganisms that evolved more recently, they should be able to achieve the same results thousands of times faster.

We could also modify them to speed up the process even further, and produce the exact results we're looking for. And we could (literally) prepare the ground for them by sending machines to mine chemicals, process the soil, and tweak the atmosphere to create optimal conditions for them to work in.

The next set of permanent colonists will probably be mechanical: orbiting spacecraft and satellites, self-controlled intelligent machines and robots, rovers, and artificial intelligence systems. There will be a *lot* of these. They'll find the best locations for the mines, industrial plants, farms, and habitats. They'll mine and process the materials to build these facilities; carry out the construction work; build utility pipelines, data networks, and other systems; make the habitats safe and habitable; and so on.

Telepresence robots will carry out tests and experiments, while being controlled (with a few minutes' delay) by engineers on Earth. We have basic robots and telepresence systems at the moment, but they would need to be far more capable, resilient and reliable before we could consider sending them to Mars. But they would allow us to do the same types of work that humans would do, without humans needing to go there[9-64].

The first farms and habitats will be constructed underground. As we've already seen, we won't be able to live and farm on the surface (except in domes) until terraforming is complete.

> We might want to continue living in domes even after terraforming is complete. We'll see why later.

The first human colonists

Following the first short-term research missions to Mars, human expeditioners and pioneer colonists will establish small research communities there. These will be similar to the Antarctic research stations here on Earth, but most of them will be located underground.

The first research communities will be built in caves and lava tubes, but some people might choose (or need) to live and work on the surface.

The first surface habitat will probably be modular – like the *International Space Station*. It will be sent to Mars one section at a time, and assembled on the surface by a small team of astronauts. Some of the sections will be expandable or inflatable, and they might be quite large inside. Over time, the habitat – let's call it *Mars Base One* – will become larger. And eventually there might be several Mars bases dotted around the planet.

We could also build small domes on the surface, each covering an area of soil, rock or ice that researchers could study at first hand without needing to go outside. The domes might be mobile, or easy to relocate, so they could be moved to new locations once work on each site was completed.

The first domes will almost certainly be made of conventional materials. And, like the habitats we looked at above, they might be inflatable. They'll need to be insulated against the extremes of heat and cold, pressurized, shielded against radiation, and contain most of the life-support, communications[9-65] and other facilities that their colleagues living and working underground would enjoy. The domes might also be divided into separate rooms inside.

> The first domes probably won't have the level of shielding required to withstand a powerful solar flare. The largest and most dangerous flares only occur once every two hundred years or so, so it's highly unlikely that one would strike during a research mission. But if one *did* strike, the domes' inhabitants would need to join their other colleagues and take shelter deep underground.

Life in these early colonies, whether in modular habitats or small domes on the surface, or in larger habitats underground, will not be particularly pleasant. It will be a major step down in terms of quality of life compared with living on Earth, and even compared with living and working in an Antarctic research station.

The colonies will be artificially lit and there will be little or no opportunity to go outside. A few highly trained geologists, biologists and engineers might be allowed outside occasionally – wearing pressurized suits and breathing apparatus, of course. But most personnel, including research scientists, lab technicians, analysts, medics, IT and communications specialists, support staff (and their families if they're allowed to accompany them) would have to remain inside at all times.

People who are confined indoors can develop issues such as boredom, depression, loss of concentration, lack of motivation, eyesight degradation, and high blood pressure. They would also feel disconnected from the Earth and from nature – an extreme form of homesickness[9-66].

> Virtual reality systems could provide some relief from these issues. They could give the inhabitants simulated (but highly realistic) experiences of walking in a forest, relaxing at the beach, hiking in the mountains, or whatever situation they prefer when they want to experience nature and the great outdoors. As well as seeing a realistic visual simulation, users should be able to experience the sounds, textures, smells, breeze, and other things associated with each place, Ideally, it should be indistinguishable from the real thing. It might even be possible to alter the simulated level of gravity, so that users can choose whether they want to feel as if they're on Earth, Mars, Eden, on a spacecraft, or somewhere else.

Studies of polar researchers and small groups of people living together in isolated communities have found that those who are introverted, self-reliant, and prefer their own company cope much better than more extrovert, gregarious types. Astronauts selected for the first missions to Mars[9-67] might be chosen on this basis as much as for their other skills and qualities[9-68]. But they'll also need to be able to work well as a team from time to time.

One of the problems with living in Martian habitats is that they would only host a fraction of the bacteria needed to maintain a healthy human microbiome. This issue has already occurred on the *International Space Station*. The microbiome in the human gut changes during lengthy space missions. The longest mission where this was tracked lasted for just one year, but missions to Mars will be much longer than this.

> Initially, researchers thought the changes might be caused by the higher levels of radiation in space. But experiments on Earth that simulated those levels of radiation didn't produce the same results. After ruling out other possibilities, the researchers found that the most likely cause of the changes was the space environment itself, and particularly the microgravity[9-69].

We don't know what the long-term effects on human health will be on Mars. The level of gravity is significantly lower than it is on Earth, but it's also significantly *higher* than it is on board the *International Space Station*.

In time, we will adapt and evolve to cope with it, just as we've (kind of) adapted to the higher level of gravity on Earth since we were brought here from Eden. In the short-term, astronauts and early Martian colonists might need to take probiotic supplements to prevent problems with their guts.

The thought of living on Mars might sound exciting initially. But once people get there and they realize how horrible it is, most of them will probably want to come home again. Unfortunately, that might not be an option.

While terraforming the whole of Mars could take at least 100,000 years, we might be able to "semi-terraform" specific parts of it more quickly.

We wouldn't be able to walk around outside without pressurized suits or breathing apparatus until the whole planet was terraformed, of course. Terraforming is a *global* operation not a *regional* one. But we might be able to warm specific regions, melt the ice, irrigate the land, decontaminate the soil, grow crops, carry out experiments, and search for fossilized microbes and other signs of life. This would become increasing possible as terraforming proceeded and the atmosphere and surface water began to be reinstated.

Permanent migrants from Earth will want to live and work as normally as possible – without wearing space suits. They could only do this if we enclosed each region inside a sealed dome.

As we noted earlier, these domes would almost certainly not be constructed from glass, Perspex or Plexiglas, as we may have envisioned in the past. Structures of the sort of scale we're talking about here would be impossible to build from these materials, even in Mars's lower gravity.

> The largest dome-shaped structures on Earth, such as the O2 Arena in London (originally the Millennium Dome), are around 1,200 feet (365 meters) in diameter. That's large enough to enclose a village or a small farm, but not a city.

Instead, we might use newer materials such as silica aerogel. It would be relatively easy to construct massive domes out of this material, and their walls would only need to be about an inch (25 mm) thick. A recent study found that silica aerogel of this thickness would cause a greenhouse effect inside the dome, which would raise the temperature by around 90°F (32°C). If the domes were located in the right places on Mars – particularly on or near the equator – they could be relatively pleasant places to live[9-70].

Again, it should be possible to automate most of the construction work. The material could be mined on Mars –

there are plenty of silica deposits – and automated systems could turn it into large-scale panels. Intelligent machines could fit the panels into place and seal them together to form each dome. Constructing the domes on Mars will be much easier than it would be on Earth, as the gravity is so much lower. But in any case, the panels will weigh very little, as they are ninety-seven percent air.

> The automated systems we develop to build these domes – and the cities inside them – could also build domes and cities on Earth. These could have a wide variety of uses, including rapidly rehousing people made homeless by natural disasters or war.

Permanent colonists are unlikely to move into these domed cities for tens of thousands of years. In the meantime, the first expeditioners and researchers will live in underground cave networks and lava tubes, sealed against the outside environment. Airlocks at the entrances and exits will preserve the atmosphere inside and keep out toxic gas and dust. The underground habitats will be fully equipped with life-support systems, oxygen recyclers, carbon dioxide scrubbers, water and waste recycling systems, heating, lighting, communications and entertainment systems, hydroponic farms, leisure and fitness facilities, and so on.

The inhabitants might be able to pump water from an underground reservoir and purify it, or use meltwater from an ice cap. They'll have electrical generators, probably with nuclear power packs as we saw earlier. And they'll only need to don their spacesuits on the (very) rare occasions when they need to venture outside.

> Robotic devices, rovers and other intelligent machines will carry out most of the outside work. They'll drill and mine samples of rock and ice and return them to the cave network where the human scientists will be able to analyze them in safety and comfort.

In the early years, the expeditioners and colonists will rely on supplies from Earth. These will arrive in large shipments two or three times a year. But once the domed cities are constructed and their environments stabilized; and mining, industrial plants and manufacturing plants are created; the Martian communities might become entirely self-sufficient.

That doesn't mean visits to and from the Earth will stop, of course. Although most exchanges between the two planets will be electronic, some goods, minerals and other materials will be traded, just as they are between nations on Earth.

Eventually, there might be more farm domes than city domes, especially if they're reasonably cheap to build and they can be fully automated. However, a great deal of the agriculture will continue to take place underground.

Once terraforming is fully complete, the colonists will be able to live and work in the open air, just as they do on Earth. There will be no need for them to live underground or in domes. They should find that living on Mars suits them extremely well by that point, as it should be *much* better than living on Earth. The lower gravity will be more to their liking, the days will be slightly longer, and the sunlight less dazzling and damaging. There will be far fewer natural disasters too. (Probably none, in fact). It won't be perfect though.

They'll still have to contend with the five-and-a-half-month winters.

With advances in space transport, it will eventually become possible for people to travel between the two planets quickly, safely, economically, and with little or no environmental impact. The trip might be considered no more arduous than traveling between the USA and Europe is today.

> While visitors from Earth should enjoy their visits to the terraformed Mars, Martians probably won't enjoy their visits to Earth. They'll find the gravity and brighter light extremely uncomfortable, and possibly dangerous. The air pressure will also be significantly higher than they're used to on Mars, and the air might also be a lot drier.

> A radical solution to terraforming Mars would be to create a new type of human that wouldn't need it to be terraformed at all[9-71]. The new species would be able to withstand the high radiation levels and freezing temperatures, and might not even need to breathe – or at least not in the conventional sense[9-72][9-73].
>
> We've seen that *Deinococcus* bacteria and tardigrades can survive in space. They can withstand the lack of oxygen, high levels of radiation, ultraviolet light, extreme temperatures, and so on. We might be able to splice the relevant genes from those organisms into our own DNA and give ourselves the same abilities[9-74].

> Creating such a species would take an immense amount of ingenuity. It would also be an ethical nightmare, and it might not even be possible. But we might be able to work around some of the trickier biological challenges using technology and electronics. The new human species might be partly biological and partly mechanical: or in other words, a cyborg.
>
> Dr. Chris Impey, an astronomer from the University of Arizona in the USA, believes the people of Mars might quickly become cyborgs anyway, even if they're normal humans when they leave the Earth.
>
> Speaking to NBC News, he said, "They will probably be aggressive in genetic engineering and self-modification, to the extent of embedding various monitoring and repair devices, and taking a cyborg path. This will be a very technology-forward cohort, advancing far beyond the average terrestrial society."

Food production on Mars

We've looked at the types of crops that would grow best in the Martian soil, and those that could be grown hydroponically in domes and tunnels. The technology to do this already exists, and all of the equipment could be sent to Mars using our current technology. If we send a manned spacecraft to Mars, a complete hydroponic system to feed the crew throughout their entire mission would take up about one-tenth of the total payload – which is perfectly reasonable[9-75]. The system would be scalable too, whether the crew was made up of just a handful of people, or thousands of them.

Even if Mars had a population of one million people, they would still be able to grow enough food using this system[9-76][9-77]. But we would need to carry out a substantial amount of preparatory work before that number of people could live there. Researchers estimate that a hydroponic system of that scale would fill nine thousand miles of tunnels. But, once again, intelligent, self-reliant machines could excavate, seal and pressurize the tunnels; install the lighting, heating and irrigation systems; plant the first crops, maintain them, and even harvest them before the colonists arrived.

A system like this would enable the first one million inhabitants of Mars to become self-sufficient within one hundred years. But in the interim they would need regular supplies from Earth. A colony of that size would need one or two shipments *every day* – or around 50,000 shipments in total. That certainly isn't feasible with our current level of technology.

It might not be possible to grow trees on Mars, and certainly not large ones. On Earth, forests are often called the "lungs of the world," as they pull carbon dioxide from the atmosphere and produce oxygen. We might not have this luxury on Mars, and we might have to create a global network of industrial plants to do the same job. These might be like giant-sized versions of the carbon dioxide scrubbers we install on spacecraft and submarines.

> The main reason why trees might not grow on Mars is because it receives significantly less sunlight than the Earth. But if we boosted the sunlight using orbiting mirrors or artificial lighting, as we saw earlier, it should become possible. They might even grow much taller on Mars because of the lower level of gravity.

An alternative would be to create billions of artificial trees. Early versions of these have already been developed on Earth. We could "plant" them wherever we would plant real trees, and they might look just like real ones and perform the same function.

Another alternative might be to fill the seas and lakes with algae. As well as soaking up carbon dioxide and producing oxygen, the algae could be harvested and processed into food, eaten by fish (which we could then eat), or turned into biofuel.

People's diets on Mars will be largely (or perhaps entirely) plant-based. As we discussed earlier, we are physiologically herbivores; we aren't designed to eat meat at all. Our bodies can't process it properly: our stomach acid is too weak to digest it promptly and it stays in our systems for too long. As it breaks down and decomposes inside us, it produces toxic byproducts. These toxins may be one of the main causes of our many chronic illnesses.

Plant-based diets of the past would have lacked protein and carbohydrates. As a result, our intelligence and the size of our brains might have been limited when we lived on Eden. We might also have lacked energy and speed. We overcame both of these issues when we arrived on Earth and started eating meat.

Plants that are packed with protein and carbohydrates are readily available now, and a vegetarian diet is perfectly feasible on Earth[9-78]. The same will be true on Mars, though with less variety on offer.

Plant-based meat substitutes are becoming widely available too, and they are increasingly indistinguishable from the real thing. Before long we might prefer them to real meat and stop eating meat altogether.

At the same time, researchers developing lab-grown meat are making significant progress. As a result, traditional animal farming could die out on Earth within the next few decades, and there should be no need for it at all on Mars.

Initially, we'll probably focus on producing energy-dense foods on Mars, rather than water-dense ones like fruit, vegetables and salad crops that take up a significant amount of space. Meat-based protein – if it's needed at all – could come from farmed insects.

> Insects are small, highly nutritious and energy-dense. And they also have useful benefits in horticulture. They can be dried and turned into protein-rich flour, which can be used in all sorts of food products. Insect-based flour, including cricket flour, is already widely available on Earth.

The range of foods available on Mars will be more limited than on Earth, at least for the first hundred years or so. It will also be more natural and less processed, probably replicating the situation that existed on Eden when we left it hundreds of thousands of years ago.

Most animals on Earth eat a limited range of food, and many eat just a single type. We would become more like them in this respect. And we might no longer be the "oddballs" of the animal kingdom.

> The main difference between the food on Mars and on our original home planet Eden is that Martian food will be more nutritious and more widely available. Our appearance will undoubtedly change as we adapt to the lower Martian gravity, and we might start to look more like we did when we lived on Eden. But we shouldn't lose any of our intelligence.

> The first food producers on Mars will focus on producing energy-dense food. Their efforts will undoubtedly lead to nutritional improvements on Earth too. It could (and *should*) even lead to the eradication of world hunger.

Traveling to and from Mars

As we saw earlier, we can send astronauts to Mars in about nine months if the two planets are at their closest points in their respective orbits. The first round-trip manned missions will probably last for just under three years. The two planets' relative orbital positions will be the main determining factor in deciding when to launch the mission and when the astronauts should return.

We could actually reach Mars in as little as four months using our current technology, if the two planets were reasonably close. But it would take a great deal of fuel to achieve the required speed. Mission controllers are unlikely to choose this option (except perhaps for the first mission), as it would be too costly and inefficient.

When we talk about cost and speed, we are of course talking about using conventional fuel and propulsion systems. By the time we're ready to make regular trips to Mars, we should have developed much faster and more efficient systems.

At the time of writing, NASA and its allies and contractors are testing a variety of propulsion systems. Many of them are classified, but one of the well-known ones is a radio frequency resonant cavity thruster – more commonly known as an EmDrive. Some researchers have hailed the EmDrive as "possibly one of the biggest breakthroughs in the history of space exploration," while others say it's "impossible" and will never work. The results from initial tests seem promising, but we'll have to wait and see what the final outcome is.

Even if the EmDrive doesn't live up to its claims, we will eventually develop other systems that can achieve the same results or better. Impulse drives or nuclear-powered rockets could reduce the journey time to as little as two weeks. A photonic propulsion system could reduce it to just seventy-two hours[9-79]. But these systems wouldn't be suitable for transporting people (or animals): the acceleration and G-forces would be too great for them to withstand.

The rate of acceleration as the spacecraft leaves the Earth and the rate of deceleration as it reaches Mars would need to be not just survivable but reasonably comfortable, and this will extend the journey time significantly. In fact, it might take as long (or longer) to travel the first and last million miles than to travel the thirty-two million miles in between.

> Let's play around with this idea. We can travel comfortably from the Earth to the Moon in three days. That's a distance of about 250,000 miles. A journey of a million miles at the same speed would take about twelve days.
>
> Now let's imagine that we were accelerating at a comfortable (and imperceptible) rate throughout that time. And the same process would happen in reverse as we approached our destination. It would therefore take twenty-four days to complete the acceleration and deceleration stages of the journey, each spanning one million miles.
>
> We could travel the other thirty-two million miles to Mars at full speed – which would also take about twenty-four days. The total traveling time for a comfortable journey to Mars would therefore be about forty-eight days. But we'll need to test this properly once we've developed propulsion systems that can achieve that kind of speed.

Unmanned craft carrying supplies and machinery could travel to and from Mars much more quickly, of course. But they (and their cargo) would need to be designed to withstand the high G-forces.

A spacecraft carrying human (or animal) passengers would also need to be shielded against radiation. The lunar missions of the 1960s and 1970s had *some* shielding, but the astronauts were only in space for a few days. The first Martian astronauts might be in space for two or three years, so their craft will need much heavier shielding.

> Astronauts and cosmonauts have spent more than a year at a time on board the *International Space Station* (ISS) without suffering any ill effects from radiation. The ISS doesn't need heavy shielding as it's in a low orbit 250 miles above the Earth's surface and it's shielded by the magnetosphere. The radiation doesn't reach dangerous levels until 400 miles above the Earth's surface.

Our current spacecraft don't have the level of shielding required for a journey to Mars, and we don't have the technology to equip them with it. We will need to develop much better shielding before we can even consider sending live passengers there. Developing this technology might take us several decades, or even longer.

> Elon Musk, the head of SpaceX, has said he plans to send humans to Mars within the next few years. As we don't have the level of radiation shielding required, this mission is unlikely to take place.
>
> The only way it could go ahead is if the astronauts agreed to be exposed to hazardous levels of radiation for the eighteen months they would spend on the outward and return journeys.
>
> This would have a significant impact on their long-term health and lifespans. One researcher said he believed all of the astronauts would develop cancer in multiple organs, as well as cardiovascular disease, and dementia[9-80], and the mission would knock fifteen to twenty years off their lives.

> Interestingly, while they were on the surface of Mars, the astronauts would be comparatively safe. Radiation levels there are roughly the same as they are on board the *International Space Station* (except for a few hot spots). The level is still significantly higher than it is on Earth, but researchers believe it falls within safe and acceptable limits.
>
> The most serious problem would come from coronal mass ejections (solar flares), which can deliver massive bursts of radiation. On Earth, the magnetosphere deflects most of the energy away, but Mars doesn't (currently) have a magnetosphere, so it would take a full hit. The astronauts would need to shelter deep underground – if they could find a suitable place and reach it in time. But even then, they might receive a hazardous and potentially lethal dose of radiation.

Issues with living on Mars

Living on Mars wouldn't be particularly pleasant until we've constructed domed cities on the surface. The inhabitants would then be able to live comfortably and move around freely within their domes.

Some people have suggested skipping the domed cities stage. They say we should jump straight from living in underground bases to fully terraforming the planet. I think that's unlikely to happen though. We've already seen that fully terraforming Mars could take at least 100,000 years. We've also seen that the weather on the surface could be horrible: it will probably be foggy, and it will rain most of the time. People will want to live inside domes even after terraforming is complete.

As we colonize Mars, I believe we'll start off with small, temporary habitats on the surface: probably expanded or repurposed landing craft in the first instance, with additional modules added later. We'll then establish larger, semi-permanent bases underground. Hundreds of years later, we might start building small domed villages on the surface, and we'll gradually build larger domed communities. After several thousand years, there might be several large cities.

By the time we get to this stage, we should have developed all of the technology we'll need to begin terraforming Mars. We should also have overcome the political, financial and ecological issues – especially if people have realized that Mars will *have* to become our new home in the future if we are to survive as a species. We'll probably begin the terraforming process while people continue to live in the domed cities.

> We will need to choose the domes' locations carefully. For example, once terraforming gets underway, the water levels will gradually rise. We wouldn't want to build communities in former oceans, seas or lakes, because they'll fill up again.

> As we saw earlier, one way to give Mars a nitrogen-rich atmosphere and more surface water might be to crash a comet or meteorite into it. The whole planet would have to be evacuated if we did this. It might remain uninhabitable for thousands of years afterward until everything had stabilized.

Even when Mars has been fully terraformed and we can live there more or less normally, people migrating from Earth will face all sorts of issues.

The lower gravity will be great for our bodies. But it will only be great in the short term (if we visit for a few weeks or months), or in the long term (as our bodies adapt and evolve to cope with the new environment). Anyone planning a medium-length stay (several months to several years) will suffer from muscle atrophy and loss of bone density, even if they follow a vigorous exercise regimen.

If they only stay for a few months, they should recover from this when they return to Earth. But they'll experience months of discomfort while their bones and muscles regain their former mass. If they remain on Mars any longer than this, the changes could become irreversible. They might suffer from major health issues or even become disabled if they returned to Earth, as their bodies wouldn't be able to cope with the higher gravity. Their muscles wouldn't be able to support them, their bones would break easily, and they wouldn't be able to walk without pain or lift heavy objects.

> Some of them might try to get back to Mars as quickly as possible (even if they hated the place) just so they could escape from the pain.

Hopefully, by the time we're ready to send large groups of people to Mars, we will have developed pharmaceutical, genetic, or other remedies to overcome this issue.

Although we must have evolved on a smaller planet than the Earth, we've lived here for hundreds of thousands of years. In that time, we've evolved numerous mechanisms to cope with the higher level of gravity, including our large, dense bones and heavy musculature. When we migrate to Mars, we'll need to *de-evolve* many of these adaptations.

Astronauts on board the *International Space Station* have noted that the process begins almost immediately. That might not be to our advantage on Mars. We'll still be large, cumbersome beasts, but we'll have weaker bones and muscles than we need – even with Mars's lower gravity.

It might take us 100,000 years or more to return to our original size and stature. Once we reach that stage, we'll be taller; we'll have thinner, lighter bones and muscles; we'll be fast and agile; we'll be able to give birth easily; and so on.

Children who are born on Mars will adapt to living there far better than those who migrate from Earth. They won't have experienced life on Earth, and their bodies will be acclimatized to the Martian environment right from birth. However, it will take hundreds of generations – perhaps as long as several thousand years – for them to acclimatize fully.

As we've seen, in the early years of Mars colonization, babies might suffer from birth defects such as encephalitis that lasts into adulthood. Many of them might not survive that long, and the infant mortality rate will almost certainly be far higher than it is on Earth.

> Again, we might be able to mitigate these issues using drugs, genetics, or other technologies.

Children born on Mars might never be able to visit the Earth. If they did, they wouldn't be able to walk very far without becoming exhausted. Not only would the gravity be too high for them, but the air would be heavier and denser. It would drain their energy and make breathing difficult.

> We might be able to overcome this issue too, using drugs, genetic remedies, and physical regimes. People born on Mars might need to take a course of medication to prepare their bodies for their visit to Earth. The medication might cause their bones to thicken, increase and strengthen their muscles, and reduce their sensitivity to pain. They might also spend time in "Earth experience centers," where the higher levels of gravity, air pressure, and sunlight will be simulated.

Over time, people living on Mars will develop different body shapes from those who remain on Earth. Eventually, they'll become a separate sub-species, or even a completely different species. We'll come back to this later in the book.

It might take us thousands of years to develop a massive space-based dipole electromagnet (or superconducting rings) to divert solar wind and radiation around Mars. It will probably take us just as long to develop an energy system capable of powering it. Until we do, the Martian colonists will need to live underground or in small shielded communities on the surface.

Once the magnet is in place and activated, we'll be able to build larger structures on the surface. They'll no longer need to be shielded, but they'll still need to be sealed and pressurized because the negligible amount of air outside won't be breathable.

The domed communities – which will eventually develop into large cities – will almost certainly exist for tens of thousands of years. They'll be supported by farms enclosed in neighboring domes or underground caves and tunnels. Industrial areas might be housed in separate domes, away from the residential/shopping/leisure domes, for improved safety and to prevent any risk of pollution or contamination. The domes might be linked together by an underground railroad network or hyperloop.

Terraforming Mars will be extremely difficult. But as we've seen, we could solve some of the hardest challenges by diverting a water-and-ammonia/nitrogen-rich comet or meteorite and crashing it into Mars. Of course, we have no idea how long it might take us to find the right sort of comet or meteorite, divert it, and crash it, nor how long it might take the planet to recover from the impact afterward.

The recovery could take millions of years, depending on the size of the comet or meteorite, which part of the planet it hit, how much damage it caused, and so on. Even if everything went to plan, we have no idea how long it might take for the ammonia to turn into nitrogen (with or without our help), what would happen to the water, or whether any of these ideas will even work.

But even if it takes us millions of years, I'm convinced we'll get the job done in the end. Mars *will* become habitable one day (in the far distant future), and its inhabitants – our descendants – will no longer need to live underground. Depending on the climate after terraforming, they might or might not decide to continue living in domes on the surface.

That doesn't mean the Martian environment will be the same as it on Earth, though. It definitely won't be.

As Mars is further away from the Sun, we'll need to give it a denser atmosphere than the Earth to trap the smaller amount of solar energy available. As we've seen, the easiest, safest and most environmentally friendly way of doing this would be to use water vapor, effectively blanketing the entire planet with clouds.

> Something similar happened on Venus. The whole planet is shrouded in dense clouds that insulate and obscure the surface. That, combined with the fact that Venus is much closer to the Sun, means the average surface temperature there is 864°F (462°C) – which is hot enough to melt lead. We obviously won't be going to such an extreme on Mars. But it does mean that people living there might never get to see the stars and other planets – including the Earth.

> Like Mars and the early Earth, Venus's atmosphere is mostly carbon dioxide (96.5%). There's also a small amount of nitrogen (3.5%), and trace amounts of sulfur dioxide and carbon monoxide.

> The aliens that brought us to Earth might have had this in mind if they tried terraforming Mars themselves before they brought us here. A thick blanket of clouds would have prevented us from seeing anything in space. We would have had no idea that stars, planets and other civilizations might exist. Of course, once we became a technological race, it wouldn't have taken us long to find out what was above the clouds.

The future inhabitants of Mars will know there are stars and planets above the clouds, of course. They'll also be able to track their movements using software, or see them in planetariums or projected onto their walls and ceilings (or the inside of their domes and caves). But they might find it frustrating not to be able to see them in real life.

The first migrants to Mars might find the permanent cloud cover and moist air depressing. It will probably rain a lot too. Later generations might come to accept it. But it certainly won't be a place for sunbathing and beach vacations.

As we've seen, the Martian citizens, might choose to live in domed cities even after terraforming is complete. They'll be able to live in dry, climate-controlled conditions that will be much more pleasant than the awful weather outside. There might also be vacation domes where the temperature is tropical and the weather is always dry and sunny – thanks to the high-power lighting.

If people can't cope with the longer seasons on Mars, the climate and light levels inside the domes could be adjusted to replicate the Earth's shorter seasons. The seasons could even be abolished completely, and sunrise and sunset could take place at exactly the same time every day, replicating the situation on Eden.

As we've seen, on some parts of Mars, the radiation levels are equivalent to the level on board the *International Space Station* and are well within safe limits. However, away from the equator, the level rises significantly, and there are radiation hot spots all over the planet.

Radiation can cause sickness, burns, DNA damage, cardiovascular damage, and cancer. It can also reduce cognitive function and fertility. People living on Mars, and those paying extended visits, will be exposed to higher levels of radiation than those living on Earth, even if they remain underground or in shielded bases. We don't know what the long-term effects might be. The Martian citizens might develop complex health issues in their later years, and some of the first generations of children might have genetic issues.

Later generations should evolve ways of coping with this. For example, their skin will probably darken and it might also become thicker. Eventually, they might even develop a different form of DNA that contains backup copies of their genes in case their primary ones are damaged. Some plants and animals on Earth have this feature.

Alternatively, geneticists might find a way of doing this using technology. For example, they might take a copy of each baby's genome when it's born, and restore it when the level of damage rises above a certain threshold. The Martians might need to visit their family geneticist as often as once a decade to have their DNA restored. As well as wiping out any damage and deterioration caused by radiation, this will also remove degradation caused by age, illness, and so on.

> If this works, it's likely that the same practice will be adopted on Earth. As the risk of radiation damage on Earth is much lower, people might only visit a geneticist two or three times in their

> lifetime – perhaps in their thirties, sixties and eighties. Those who are more prone to genetic issues might need to visit their geneticist more often. This should solve many of the health issues that arise from DNA degradation in later life.

> Eventually, geneticists should be able to fix *any* genetic health issue by editing strands of DNA. For example, if the geneticists determined that someone was at risk of developing cancer in their later years, they could remove that weakness.

Another issue the first colonists will have to contend with is the sheer distance between Mars and the Earth. There will be communications delays of at least three minutes and sometimes as long as twenty-two minutes, depending on the planets' orbital positions. Real-time communication – including phone and video calls – will be impossible. Unfortunately, there's (probably) no way around this issue, whatever technology we develop in the future. Data can't travel faster than the speed of light.

> I'm not ruling this out completely, though. Researchers are working on ways of allowing spacecraft to traverse wormholes that link one part of the universe with another. The spacecraft would be able to take a shortcut through the wormhole and arrive in another part of the universe in a fraction of the time it would take to fly there the conventional way. They will "appear" to have traveled faster than light, without breaking any

laws of physics. While the researchers are working on ways of achieving this with physical objects, they might find a way of sending data (in the form of photons, electrons, or something else) back and forth through the wormholes. If there was a permanent data wormhole between the Earth and Mars, near-real-time communication *might* become possible.

Other researchers are working on quantum entanglement at increasingly large distances – and they appear to be having a degree of success. If they can make it work between planets, real-time communication between the Earth and Mars – or between *any* two points in the universe – could eventually become possible.

It might take us hundreds of thousands of years to crack these problems. But that's fine, because it will probably take us that long to develop the other technologies we'll need.

There could be even longer delays in communications if the Earth and Mars were on opposite sides of the Sun. The Sun would block the flow of data, causing blackouts that could last between two weeks and a month. But this could be mitigated easily using regular communications satellites.

Anyone visiting Mars, or living there in the next several thousand years, will need to get used to the idea that all communications with the Earth will be delayed. It will take several minutes for a message from Mars to reach the Earth and several minutes for the response to come back.

> Real-time communications wouldn't be possible (at least in the short term), but audio-visual communications *would* be possible. The sender could record a message and send it to the recipient, who would receive it several minutes later. The recipient could then record a reply and send it back. The conversations wouldn't be as instant and spontaneous as they are on Earth, but people would still be able to talk to each other and work together. They might even have long-distance relationships, even if they never meet in person.

People on Mars could also receive television and radio broadcasts from Earth. And, of course, people on Earth could receive broadcasts from Mars. Again, there would be several minutes' delay while the signal crossed space. But Martian media companies could record broadcasts from Earth and rebroadcast them on their own networks whenever they liked.

> As the Martians evolve and their appearance changes, the people of Earth will start to look like a different species. The two races will lose the sense of connection they once had with each other.

9. Making Mars Habitable

> Imagine if we were able to watch alien television broadcasts now. The aliens would look strange, and their television shows might be beyond our comprehension. We might experience the same thing when we watch Martian broadcasts in 100,000 years' time. And the Martians might experience the same thing when they watch *our* broadcasts. But *some* of their broadcasts might make compelling and unmissable viewing!

The first colonists on Mars will probably suffer from separation issues – a severe form of homesickness. They might never be able to return to Earth again. And that would mean a complete separation from their places of birth, their family and friends, Earth's flora and fauna, and everything else.

> As we've already seen, virtual reality systems could provide some relief from this. While that should work well for those colonists who were born on Earth, it might not work so well for those who are born on Mars. The Earth will seem like a weird place to them, and although they might enjoy visiting it in virtual reality, it won't mean as much to them. Similarly, they might enjoy exchanging messages with their grandparents on Earth, but they might never meet them in person. So they might not develop the same bond that they would have done on Earth.

> Future generations of colonists who are born on Mars won't experience the separation issues that their parents suffer, and they'll accept their living conditions as "normal." But they *might* suffer from severe jealousy when they see how the people of Earth live: able to go outside, walk in the sunshine, visit beaches and forests, shop in city malls, and so on. They won't be able to do any of those things on Mars. They'll be living in small communities, mostly underground, and they won't be able to go outside. Even if they *were* to go outside, there would be nothing to see but rocks and dust.

> The night sky should look spectacular from Mars though. There would be no light pollution, and the thin atmosphere wouldn't cause the distortion and shimmering effects we get on Earth.

Some colonists will feel that living on Mars is like being in prison*, and they might not be able to cope. They might have mental breakdowns and have to be moved to special medical facilities – like asylums or mental hospitals. They might have to spend time living in a virtual world – a realistic simulation of Earth. Some of them might even have to be brought back to Earth. But, as we saw earlier, depending on how long they'd been on Mars, they might then suffer from physical issues caused by the higher gravity, air pressure and solar radiation, and wish they were back on Mars again. But if they'd been returned to Earth for medical reasons, the authorities would be unlikely to approve such a request.

> *You might like to refer back to our earlier discussion about the Earth (or Mars) being a prison planet.

Children who are born on Mars and visit the Earth (after undergoing the preparatory drug and physical training regime we discussed earlier) might realize how much better life is there, and they might not want to return to Mars. They might go missing or defect.

> On the other hand, when they discover there's no escape from the higher gravity, and the sunlight dazzles them and burns their skin and hurts their eyes, and the air is twice as heavy as they're used to, and it's dry and difficult to breathe, and they feel permanently exhausted ... they might not want to stay on Earth for very long.

Another issue the first waves of expeditioners and colonists will have to deal with is the fact that they'll be living in close proximity with other people.

The first few groups of astronauts will have the biggest problem. Let's say there are a dozen people on each of the first few missions. Think about the twelve most obnoxious people you've ever worked with, and imagine having to spend three years living in a small house with them, with no way of getting away from them. You might start out as good friends, trust each other, and appreciate each other's skills, but it's unlikely things will remain that way for the entire three-year mission.

> I wouldn't be surprised if someone was murdered on Mars – or on the journey to or from it – during one of the first missions.

> This leads us to an important point: people will die on Mars – hopefully from natural causes after enjoying long and happy lives. But what should we do with their bodies? We *could* bury them or cremate them as we do on Earth. But, given the scarcity of resources on Mars, some researchers believe we should recycle them[9-81]. Their system would break the bodies down into raw materials (including compost) that could then be reused.

Astronauts who are selected for these missions – and those applying to live permanently on Mars – would need to undergo extensive psychological screening. We saw earlier that introverts who prefer their own company would have an advantage, but they would also struggle with living and working so closely with the other members of their team or community. Once again, virtual reality systems could provide a solution here. They could make users feel as if they're living and working in much larger spaces than they actually are.

Sport on Mars

As a brief aside, let's consider sport on Mars. Martian athletes will be able to jump higher and longer; run faster; kick, hit or throw balls and other objects much further than anyone can on Earth; and so on. Sports fields and stadiums will need to be significantly larger, and just about every sporting record on Earth will be smashed on the first day of competition.

But athletes from Earth and Mars would never be able to compete against each other. Terrestrial athletes who traveled to Mars would have an unfair advantage over the Martians, being bigger and stronger and used to higher levels of gravity. They would also have a huge advantage over any Martian athletes who traveled to Earth, because the Martians would hardly be able to run or jump at all.

Equipment needed for colonization

Along with their habitats and life support systems, the first colonists on Mars will need a wide range of machinery, equipment and tools. Most of this would be shipped to Mars, installed, tested, and set into operation before they arrived.

Let's consider their habitats and workplaces first, and assume they'll be set up in networks of caves and tunnels underground. The first task will be to establish exactly how far underground they will need to be. We'll need to send detecting equipment into potential habitats to check the background level of radiation, and their ability to withstand energy bursts from solar flares. The levels will be different for each site, depending on its depth, the type of rock, its longitude (the angle of the site in relation to the Sun) and so on. That means each site will need to be checked individually. In fact, each site might need to be checked several times to see how the level varies over the course of a Martian year.

> Rather than repeatedly checking each site, it might be more efficient to install monitoring devices in potentially habitable sites. The devices would send data wirelessly to a receiving station or satellite.

The first colonists could live in natural caves and lava tubes. But the caves and tubes would almost certainly have to be enlarged to accommodate larger groups as more people arrived from Earth.

> Once terraforming gets underway, the air pressure on the surface of Mars will begin to rise. We will need to take this into account, because carbon dioxide could be forced down into the underground habitats.

As we saw earlier, we could install hydroponic farms in tunnels underground. Each tunnel would be several miles long and around twelve feet in diameter. We might be able to use some natural tunnels, but most of them would need to be excavated. They would also need to be fitted with air supplies and lighting, along with supply lines for water and nutrients. If people would be working in them, they would need breathable atmospheres and heating. But if they were unmanned or operated by robots, their atmospheres and temperatures could be tuned for optimal plant growth and cost efficiency, rather than for human safety and comfort.

Areas will need to be set aside for storing the harvested crops. These areas might be refrigerated or they might make use of Mars's natural low temperatures. We would also need to prevent the stored food from becoming contaminated by dust, radiation, and other pollutants.

The inhabited parts of the cave and tunnel network would need heating, sanitation and waste recycling systems, communications and entertainment systems, and medical facilities. Separate rooms will be allocated to living and working spaces, bedrooms or private apartments, bathroom facilities, kitchens, storage, recreation, and so on.

If there are many inhabitants, there might also be shops, restaurants, places of worship, theaters, and other amenities. Each of these might just be a single room to start with.

As the population increases, we would need to think about building schools, hospitals, art galleries, concert halls, libraries, sports centers, arenas and stadiums, shopping malls, and so on. These might need to be carved out of the rock underground, then sealed, pressurized and monitored – with back-up systems that would kick in instantly if anything malfunctioned.

> We won't get to this stage for tens of thousands of years. Until then, everyone will live in small research communities, like the Antarctic bases on Earth. These have a maximum of about 100 staff.

Few (or perhaps none) of the first inhabitants would consider their Martian research base a permanent home. Most of them would return to Earth after a few months or a few years at most, at the end of their "tour of duty." But their time on Mars would have to be carefully planned and monitored, so they didn't lose too much bone density or muscle mass before they were scheduled to return home.

We will need ultra-high-efficiency electrical plants to power all of these systems. These might take thousands of years to develop, as our current systems are nowhere near powerful enough, efficient enough, or reliable enough.

Nuclear fusion would be a good option, as the systems would be safe, long-lasting and compact, with widely available fuel and no radioactive waste. The main drawback is that they would need to be refueled from time to time.

Solar power is another good option, particularly as it wouldn't need to be refueled. But the photovoltaic arrays would need to be massive, and many orders of magnitude more powerful and efficient than our current ones. They could be placed in orbit around Mars, of course, sitting alongside the solar reflecting mirrors we looked at earlier.

But we're still talking about a time in the distant future, thousands and thousands of years from now. We will almost certainly have developed an entirely new form of energy by then – one that we can't even conceive of right now.

> The new energy systems will have huge benefits on Earth too, of course.

Every system we install on Mars will need to be robust and able to withstand the extreme environment. They'll need to cope with widely fluctuating temperatures, extreme cold, high levels of radiation and UV, dust storms, and so on. But if we can make them work on Mars, we could also use them in space exploration and the colonization of other planets.

The systems will need to be easy to service and maintain, and it would make sense to design them with robotic maintenance in mind from the outset. They could even be designed to be self-maintaining, meaning that they would automatically request and replace parts themselves if anything went wrong. In the rare case that human engineers needed to work on them, they should be designed so that as much of the work as possible could be carried out indoors.

> Indoors might mean underground, in a maintenance dome on the surface, or (in the case of the dipole magnet in space) inside a spacecraft.

One of the biggest maintenance operations in the early years of Martian colonization will be the removal of dust. Martian dust is abundant, as fine as talc, gets into everything, and covers everything, reducing its efficiency and eventually putting it out of operation[9-82]. Automated systems should be able to remove it, minimizing the length of time human engineers would need to spend outside.

> Martian dust has put several of our rovers and landers out of action — mainly by covering their solar panels and causing them to lose power. We've known about this problem for decades, so I'm not sure why the most recent vehicles haven't included dust-removal systems (such as fans or vacuums) to counteract this problem.
>
> Unfortunately, the dust (and the acidic water that used to cover the surface of Mars) may have eroded any signs of previous life.

Dust storms sometimes block out the Sun for months at a time. If the Martians rely on solar power systems, they'll also need backup systems, such as nuclear power generators, or ultra-high-capacity batteries.

The Martian colonists will want to use local water supplies as much as possible, for drinking, washing, irrigating their crops, and so on. But the local water supplies might be too salty, or contaminated with toxic chemical elements or heavy metals — particularly if they're located far from the poles. So the colonists might need water purification systems and desalination plants.

Once the colonists have completed their initial research missions, they might begin mining Mars for minerals and building materials. They'll probably construct small mining communities, and the most successful ones will almost certainly evolve into cities, just as they did on Earth. The colonists will need heavy mining machinery, and industrial plants to process the materials they extract.

Some of those materials, such as rare minerals, might be exported to Earth. If the colonists want to do this, or return to Earth themselves, they might need to produce their own fuel so they can get there. They might be able to turn some of the materials they mine into fuel, or grow crops and turn them into biofuel. Of course, they would need facilities for developing, testing, maintaining and launching spacecraft too.

Finally, the colonists will need some means of getting around on Mars. They might have electric vehicles that they can recharge using the energy systems that power their other equipment. Or they might have dual-fuel or multi-fuel vehicles that switch to other types of fuel if they're too far from a recharging point.

Potentially, the colonists could use aircraft to get around. They wouldn't be able to use conventional aircraft until after Mars was terraformed, because the atmosphere wouldn't be dense enough to support them. But they might be able to use things like helium-filled airships.

Another option, for long-distance travel, would be to launch themselves into space and orbit the planet until they reached their destination. As the gravity is lower on Mars, this would be significantly easier and cheaper to do than it is on Earth.

Conventional aircraft including helicopters, airships and hot air balloons might be able to operate inside the domed cities – as long as the domes were large enough – as they would have dense, pressurized atmospheres. These vehicles would be particularly useful for maintenance operations.

After terraforming

Once terraforming has been completed, Mars should closely resemble our original home planet Eden.

It should also resemble the Earth in many ways, but the gravity will be lower, the light will be dimmer, the air pressure will be lower, it will be wetter, the seasons will be longer, and so on. Conventional aircraft will be able to fly there – though I'm sure we will have developed better alternatives by then.

We will no longer need to live and work in domes or cave and tunnel networks, and everyone will be free to travel and live wherever they please – as long as the political leaders allow it. If they don't want to remain in the cities, the colonists will be able to establish new towns, villages, hamlets, and other communities elsewhere.

They'll be able to build conventional houses out of conventional materials if they want to, using locally sourced rock, sand, clay, and metal ores.

Once we get to this stage, the emphasis will switch from research to production, and the jobs people have on Mars will be similar (or identical) to the ones people have on Earth.

Manufacturing companies will produce goods that were previously imported from Earth, and service industries will cater for everyone's needs.

I would imagine that just about everyone will have jobs. There will be plenty of work to do, and everyone will be expected to work hard and make an active contribution to society. Productivity will be high. This is likely to be the case for several generations. And, bearing in mind the wet weather and the dim light, it wouldn't be the kind of place you would choose to retire to anyway.

> Having said that, there would be other benefits if you were elderly, sick or disabled. The lower gravity would suit people with muscular and skeletal problems, the lower air pressure would suit people with breathing issues, and the lower light would suit people with certain eye conditions. However, by the time we can live freely on Mars, we should have developed cures for those conditions anyway.

A significant amount of the work will be mechanized and automated, of course. But there will still be plenty of work for people, who will do the things that machines and artificial intelligence can't.

Let's jump ahead again to one million years in the future. The cities have been completed, and robots do all the servicing and maintenance. Mars is a pleasant, clean and safe place to live. Millions more migrants have arrived from Earth. They don't have jobs, and there's nothing for them to do. This will be the age of leisure and universal income.

> The political leaders and their intelligent machines might have to devise ways of keeping everyone occupied – on Earth as well as on Mars.

The Earth should also be a clean and safe place to live by then. Machines, robots, and artificial intelligence systems will have cleaned up the land, oceans and atmosphere and removed all the space junk. Advances in genetics will have tamed our innate violence and, combined with smart materials and buildings interwoven with sensors, put an end to things like anti-social behavior, drink- and drug-related issues, littering, vandalism, graffiti, shoplifting, mugging, robbery, and other forms of crime. Smart buildings, construction materials, fabrics and electronics will have built in fire-proofing. Robots will patrol the streets, waterways and skies, and keep the beaches safe. All industrial processes, including power and heat generation will be safe, clean and energy-efficient. And all travel will be clean, safe, either free or very cheap, and harmless to the environment.

Is it worth the effort?

The tremendous cost, time, and effort involved in terraforming Mars might make us question whether it would even be worth doing.

If we really wanted (or needed) to live on Mars, it would be far easier – and faster – to cover it with domes and build cities and farms inside them. If we did that, we wouldn't need to develop the ultra-efficient energy systems, put giant magnets in space, crash a comet or asteroid into Mars, wait hundreds of thousands of years for nature to take its course, or do most of the other things we've looked at in this chapter.

Building domed cities on Mars is way beyond our current level of technology, of course, and it will probably be thousands of years before we build the first one. But once we've covered Mars in them, and created a transport network to link them together, it might make sense to stop there and forget all about terraforming it.

We could tune the atmospheres inside the domes to perfectly suit our physiologies. And, as we saw earlier, the domes would be much nicer places to live anyway: the weather outside would be cold, cloudy, damp and horrible. We could replicate the Earth's seasons inside the domes, so they only last for three months rather than nearly six, or we could do away with them altogether. We could also increase the light levels, fill some of the domes with trees and forests, create beach areas for vacations, and a whole lot more. And we'd be able to do all of this while enjoying the lower levels of gravity and air pressure.

By the time we're able to build domed cities on another planet, robots should be able to do all of the outside work anyway. So why would anyone want to go outside?

On the other hand, the technologies we would develop as we worked toward terraforming Mars would have untold benefits here on Earth. We'd probably find ways of solving the climate change problem and preventing comets and asteroids from hitting us. We'd develop fantastic new systems for power generation, agriculture, transport, construction, medicine, genetics, artificial intelligence, and hundreds of other things that would make dramatic improvements to our lives.

But the cost of doing all of this would be astronomical, there would be political and environmental objections, centuries of legal wrangling, and the whole project would take so long that most people would lose interest. Many of the greatest projects we've accomplished were started and then abandoned, before eventually being restarted and completed. Some of them were restarted and abandoned several times over. I can see this happening on Mars too.

So, is it worth the effort? Should we even bother? Maybe. Maybe not.

> NASA engineer Adam Steltzner came up with a term known as the *terraforming paradox*: the skills and abilities required to change another planet to suit our needs are the same as those required to keep the Earth habitable and sustainable. He said: "We won't be able to get the job done on other planets until we figure out how to do it here."

But, one way or another, we will all have to migrate to Mars within the next 600 – 800 million years (or less) because the Earth will become uninhabitable.

Alternative: terraforming Venus [9-83]

Some researchers have suggested that terraforming Venus might be a more practical option than terraforming Mars. I really don't think it would. While some of the processes would be similar, Venus also presents us with some very different challenges.

The main advantages would be:

- Venus's orbit is closer to the Earth than Mars, and when the two planets are aligned, the traveling time between them is shorter. If it takes us nine months to travel to Mars, it would only take us six months to travel to Venus.

- Venus already has a thick carbon dioxide-rich atmosphere, and the air pressure is more than high enough. In fact, it's so high that aircraft and spacecraft would float in its atmosphere, as long as they were airtight and pressurized.

- There's an abundance of solar energy.

- Unlike Mars, there's too much carbon dioxide rather than too little of it. That means there's plenty we could convert into fuel.

But on the other hand:

- Venus's atmosphere is *incredibly* dense and the air pressure would be too high for us to survive. Spacecraft are crushed by the pressure and stop working within a few minutes of landing on the surface.

- It rotates really slowly on its axis, but orbits the Sun more quickly than the Earth does. A Venusian day lasts for eight Earth months, while a Venusian year lasts for seven and a half Earth months – so its days are longer than its years.

- It lacks a magnetic field and magnetosphere, just like Mars – but for a different reason.

- As the Sun becomes hotter, Venus will become uninhabitable much sooner than Mars. The Sun might not consume the Earth when it expands into a red giant star, but it will definitely consume Venus.

But let's assume we decide to terraform Venus anyway, even though it's a terrible idea. Here's what we would need to do:

Create a magnetosphere

Venus lacks a magnetic field, so our first priority would be to either create one or divert the solar wind, harmful radiation and energy bursts from solar flares safely around the planet. We've already looked at how to do this on Mars. There are two primary options: building a system of refrigerated superconducting rings around Venus (which could also be used as an energy transfer and storage system) or placing a magnetic dipole shield between Venus and the Sun.

Remove most of the carbon dioxide

Researchers have proposed several ideas for removing the carbon dioxide, including:

- using bacteria to turn the carbon into organic molecules

- using chemistry to turn it into carbonate minerals

- adding hydrogen to convert it into water

- injecting it into the basalt rock – which is what happened to most of the water on Mars

- or cooling it to form dry ice, which could be shipped off-planet

If we could thin the atmosphere first (see below), we could grow plants on Venus, and these would take up a significant amount of the carbon dioxide.

Thin the atmosphere

Researchers have so far failed to come up with any practical solutions for reducing the density of Venus's atmosphere. It's so dense that methods that would work elsewhere, such as bombarding the planet with asteroids, would have little or no effect. One option might be to physically pump it into space – but that certainly isn't practical.

Cool the surface

Venus receives twice as much solar energy as the Earth, so we have the opposite problem that we have on Mars. We would need to cool the surface – and prevent it from heating up again. Thinning the atmosphere and removing most of the carbon dioxide would help to some extent, but it wouldn't be enough. Researchers have proposed several options, including putting giant shades or reflectors in the atmosphere (perhaps suspended from balloons), or using heat pipes to radiate heat from the surface out into space.

Add water

We could create plenty of water by adding hydrogen to the carbon dioxide that's already present. Some researchers believe there could be massive reserves of hydrogen* inside

Venus that we could use. Alternatively, we could collect it from Jupiter, Saturn or another planetary body, compress it, and ship it to Venus in massive space tankers. Creating water from carbon dioxide and hydrogen requires iron, but we could mine that on Mercury or the Earth's Moon.

> *The latest research on Oumuamua, an interstellar body that caused great excitement (and intense speculation) when it passed through our solar system in 2017, suggests it's composed of frozen hydrogen[9-84]. The research also suggests that the Milky Way galaxy could be full of similar objects. These bodies could prove to be fantastic sources of hydrogen that we could use for fuel and terraforming projects.

Other options include crashing a water-rich comet into Venus, shipping water ice from one of the moons of Jupiter or Saturn, or capturing a small, icy moon from the outer solar system and crashing it into Venus.

> We could also do this on Mars.

> The surface of Venus is extremely flat. If we added just ten percent of the water on Earth, it would cover eighty percent of the surface, even though the two planets are practically the same size.

Consider making the days shorter

We could place giant mirrors in orbit around Venus to reflect solar energy onto the areas that would otherwise be in darkness for two months at a time. Shades in orbit could also darken the parts that would have days that last for two months.

If the mirrors and shades followed a 24-hour orbit around Venus, they could replicate the length of an Earth day. But ideally, of course, we'd want to replicate the length of a day on Eden, which is 25 hours.

A less practical alternative would be to physically alter Venus's speed of rotation. This is almost certainly impossible. But it could (in theory) be achieved by causing thousands of asteroids to make close fly-bys.

> Some researchers have suggested that we could use a similar process to move the Earth into a wider orbit as the Sun grows hotter and expands. However, it might take a billion years to develop the technology to do this, which could be too late.

Having said that, there's a good argument for leaving the length of the days on Venus unchanged. It has such a dense covering of cloud that the days and nights are more like seasons that last for two months. Researchers believe that if Venus had a thinner, Earth-like atmosphere, the maximum temperature would be around 95°F (35°C) even on the side facing the Sun.

One other proposal, which solves many of these problems, would be to use robots and intelligent machines to create artificial mountains on Venus. We would build communities on top of the mountains, where the atmosphere would be thinner and the temperature cooler. The communities would be housed in domes, just as they would be on Mars, so they would have breathable atmospheres.

The biggest problem with this approach is that the mountains would need to be more than thirty miles high. The machines that build them would also need to be hardened against the intense pressure, and they would need heat shields and internal cooling systems.

Right first time

Before we begin terraforming Mars, we need to be absolutely sure the methods we're using are the right ones. Each step in the process must be logical, and prepare the way for the next step. Every step should be tested in advance using scale models and computer simulations, so we know exactly what will happen and what the exact outcome will be.

This is vital, because if we get any of the steps wrong, there will be no going back. We only get one chance, so everything has to be right first time[9-85].

In the next chapter we'll look at the many possible reasons why the aliens might have brought us to Earth.

10

Humans On Earth

Why did we end up here on Earth rather than Mars – or anywhere else?

Obviously, the aliens' original plan to take us to Mars changed when its environment collapsed. There was no chance that Mars would develop naturally into a facsimile of Eden, with vegetation and a breathable atmosphere. So why didn't they terraform it and drop us off there a few hundred thousand years later? It's unlikely that they lacked the knowledge or technology, as we've pretty much figured out how to do it ourselves. The main factors would seem to be the time, cost, or political will.

Eden would have been so much like Mars that the aliens might not have had to make any alterations to our genome to enable us to live there. But when they realized how much time, effort and money it would take to save the Martian environment from collapse, it must have given them pause for thought. They'd agreed to save us, and to relocate us to a remote part of the universe. They thought they'd found the perfect new home for us; one they could relocate us to easily and without any complications. Now they realized they hadn't.

Bringing us to Earth was the next best option. It was still in a remote part of the universe, it was right next door to Mars, it was habitable – to an extent – and its environment was stable. At first sight, the cost and ease of bringing us here would be more or less the same as taking us to Mars.

But, even though we should have been be able to survive on Earth, there were complications. The gravity and solar radiation were much stronger than we were used to. The Earth had seasons, which we weren't used to – but then so did Mars, and the ones there are nearly twice as long as the ones on Earth. There were tectonic plates, which meant there were earthquakes, tsunami, active volcanoes, and other natural disasters we couldn't detect or escape from in time. The air pressure was double what we were used to, which would multiply the effect of the higher gravity and make breathing more difficult. The oxygen level was probably slightly higher too. And while that would give us a little more energy to compensate for the higher gravity and air pressure, it would also increase our risk of dying in a fire.

As an experiment, around 300 million years ago the aliens gathered a group of us together and dropped us off on Earth to see what would happen. They probably erased our memories too, so we had no idea where we had come from or the level of technology we had once enjoyed.

But it didn't go well. The entire group died.

We don't know how long they managed to survive. We've found evidence that they were here, but not a lot of it, so they couldn't have been here for long.

At this point the aliens probably began tinkering with our genome. They identified the main issues that caused us to die, and modified the relevant genes to address those issues.

Each time they did this, they brought another group of us to Earth and watched to see what happened. Although each new group probably survived a little longer than the previous group, they all still died.

The evidence suggests that the aliens may have given up at this point. No further modern human artifacts appear in the historical record until 55 million years ago – and the people who created those artifacts died out too.

The aliens don't appear to have tried again until the native hominins evolved on Earth.

The aliens would have analyzed the hominins' genomes to see what they could learn – and more importantly, which genes they could borrow. At first, they may have spliced a few of their genes into our DNA. Sometimes it may have worked, but more often it probably didn't. But, eventually, they managed to get a group of us to survive on Earth for what they considered to be a reasonable period.

We still couldn't cope with the environment though: we probably became sick quite quickly, and we were weak and feeble. We struggled to move around, we couldn't breathe very well, the light hurt our eyes and burned our skin, there was hardly anything to eat, and the water made us ill.

The aliens had probably dulled our brains, memories and senses too, so we wouldn't kill the hominins. But that meant we were easy targets, and the hominins may have killed *us*.

The hominins might also have kidnapped and raped our women. And while that would have been a horrifying thing for the women involved, it would prove to be a massive boost for humanity as a whole. The children they gave birth to would have been hybrids – part human, part hominin – and they would have had the genes they needed to survive here.

Over time, as the hominins raped more of our women, and the resulting children had children of their own, we became stronger and better adapted to the Earth's environment.

> Eventually, we became strong enough to fight back – and we did so with a vengeance. We drove the hominins to extinction. We may have attacked them physically, we may have seized their territory, we may have out-competed them for food and other resources, and we may have given them diseases they had no immunity to.
>
> The latest research also suggests that Neanderthals and modern humans may have engaged in warfare[10-1]. We kept testing their finest fighters – and losing. But once we became strong enough, we not only stopped losing the battles, we wiped them all out and took over their territory in Europe and Asia.
>
> There's no question that we were ultimately responsible for the hominins' extinction. But a small part of them lives on in every one of us today.

> Another reason why the aliens may have dulled our brains is to extend our lifespans. Researchers have discovered that people with a reduced level of neural activity live longer than other people. In experiments, they also found that if they used drugs to suppress neural activity in nematode worms, they lived longer[10-2].

The aliens that brought us here may have added Neanderthal (and Denisovan) genes to our DNA too. Some geneticists have reported finding scars in our DNA where sections appear to have been cut out and new sections spliced in. If the aliens added those sections, they almost certainly harvested them from the hominins.

This could account entirely for the small percentage of hominin DNA that every person on Earth has in their genome today. But I believe the rape theory is also a good one.

In fact, *both* of those things could have happened. The rapes might have been unexpected, as the aliens probably thought *we* would be the ones doing the raping, not the hominins. But it saved them from having to make more extensive modifications to our genome.

> There's a chance that the rapes could have happened the other way around, of course. Human males might have raped hominin females. But as we were undoubtedly the weaker species when we first arrived on Earth, and we were struggling to survive in a harsh, unfamiliar and unforgiving environment, I think it's more likely that the hominins did the raping. The outcome was the same either way: much stronger semi-Earth-adapted human-hominin hybrids – better known as modern humans.

I don't believe the changes the aliens made to our genome were particularly extensive. They'd already given up on terraforming Mars because of the time and cost involved, so they wouldn't have put too much effort into rebuilding our genomes. So, once they'd made a few modifications to our genome and checked we could (just about) survive on Earth, they more or less abandoned us here.

I'm sure they must have been delighted when we created our own hominin-human hybrids – whichever way around it happened. It would have saved them a huge amount of effort.

> They didn't abandon us *completely*, of course. They've returned now and again to check on our progress. And now that we've entered the technological age, they're almost certainly monitoring us continuously.

There's another possibility, which I mentioned in *Humans Are Not From Earth*: humans and hominins might have mated with each other for a dare or bet.

For example, the humans – possibly drunk from eating fermented apples they'd found – might have dared each other to have sex with a hominin. Occasionally, either a human or hominin female might have become pregnant as a result.

Most of the pregnancies would have failed, and even when they resulted in a birth, the children would have been infertile. But perhaps once in every fifty years, one of the children may have survived into adulthood and been fertile. One of those children could be the ancestor of all of us. It's another possible explanation of how we all ended up with a small percentage of hominin DNA in our genomes.

> In reality, it could have been any (or all) of those things.

When we came to Earth

The aliens brought the first humans to Earth in small groups of perhaps a few hundred people – just enough to establish a breeding colony. Later, they brought much larger groups consisting of thousands of people. The evidence we've found suggests that these larger groups were brought here thousands of years apart, they were dropped off in different parts of the world, and only the most recent groups survived. We've found plenty of evidence that the groups were here, but few mainstream scientists accept the evidence or will even agree to examine it.

Here is a shortened and revised version of the timeline that first appeared in *Humans Are Not From Earth* (Second Edition).

4.82 billion years ago
Our home planet Eden formed (assumption).
This is the youngest it could be. It could be as much as a billion years older than this.

4.603 billion years ago
The Sun and Mars formed[10-3].

4.543 billion years ago
The Earth and the other planets in the solar system formed.

4.474 billion years ago
The first life-forms appeared on Eden (assumption).

4.2 billion years ago
The first life-forms appeared on Earth (pretty much confirmed, but not universally accepted).

3.5 billion years ago
The first single-celled organisms appeared on Earth.

2.45 billion years ago
The Great Oxygenation Event on Earth.

600 – 900 million years ago
The first multicellular organisms appeared on Earth.

530 million years ago
The first true vertebrates appeared on Earth.

500 million years ago
The oldest known non-natural artifact on Earth. Zinc-silver alloy embedded in a vase inside a block of coal (discovered in Massachusetts, USA in 1851). If this is artifact is genuine, it must have been left by short-term visitors (or time travelers) as there would have been nothing for them to eat on Earth. They probably weren't human, and they almost certainly didn't come from Eden, as humans hadn't evolved there yet.

465 – 470 million years ago
The first land plants appeared on Earth: mosses and liverworts.

400 million years ago
Non-natural artifact: the London Hammer[10-4] – an 1800s-era iron hammer encased in 400-million-year-old rock. It was found in London, Texas, USA in 1936. It was almost certainly formed by concretion in modern times, but I've included it here for completeness.

397 million years ago
The first four-legged animals appeared on Earth.

385 million years ago
The first trees appeared on Earth.

320 million years ago
The first hominins appeared on Eden (assumption).

313 million years ago
A species equivalent to modern humans was living on Eden (assumption).

312 million years ago
Non-natural artifact: an iron pot found inside a lump of coal in Oklahoma, USA in 1912. This pot may have been made by one of the first groups of humans to be dropped off on Earth. The group didn't survive. As humans had only recently evolved on Eden, it seems rather early for them to have been brought to Earth. So my feeling is that this is not a human artifact, but was left here by aliens.

300 million years ago
Non-natural artifact: a piece of aluminum gear was found embedded in coal in Vladivostok, Russia in 2013[10-5]. This may have been created by another group of humans that were dropped off here. Again, they didn't survive. An alternative explanation is that it may have come from a crashed alien spacecraft or a broken scientific instrument the aliens used. Mainstream scientists claim it's simply the fossil of a naturally occurring ring-shaped organism. But they are unable to explain how it came to be made from crystalized aluminum.

252 million years ago
The first Great Mass Extinction (the Great Dying). A series of volcanic eruptions wiped out ninety to ninety-six percent of life on Earth. The Earth took around 1.3 million years to fully recover, and the dinosaurs then became the dominant group of species for the next 185 million years.

200 million years ago
The second Great Mass Extinction. The Triassic period ended and the first mammals appeared.

70 million years ago
The first grasses appeared on Earth. Interestingly, grass pollen is one of the main causes of hay fever in humans. If we had evolved here 200,000 to 300,000 years ago, as mainstream anthropologists insist we did, then this would make no sense. The grass would have been here long before we were, and we should not be allergic to it.

66 million years ago
The Cretaceous-Tertiary (K/T) extinction. This was probably caused by an asteroid, which struck the Gulf of Mexico. The dinosaurs were wiped out, clearing the way for the mammals to dominate the planet.

55 million years ago
The first true primates appeared on Earth.

55 million years ago
Non-natural artifacts: stone tools, utensils, and vessels have been found embedded in rock strata beneath Table Mountain in Cape Town, South Africa. These artifacts could not have become buried in the rock due to concretion in modern

times – they are *definitely* 55 million years old. The aliens appear to have dropped off a much larger group of humans after a significant break. These people may have survived for a few generations, but they eventually died out.

25 million years ago
The first apes appeared on Earth after splitting from the Old-World monkeys.

7.2 million years ago
The ape and hominin lineages diverged.

5.8 million years ago
Possibly the first hominin on Earth to walk upright: *Orrorin tugenensis*.

2.8 – 3 million years ago
The first members of the Homo genus (*Homo naledi*) appeared on Earth, and the hominins' brain size increased significantly.

2 – 2.5 million years ago
Non-natural artifact: shark teeth with holes bored in them (probably for stringing on a necklace) were discovered in England in 1872. Analysis indicates they were created by modern humans, but their civilization didn't survive.

> The Neanderthals also bored holes in things like eagle talons and wore them as necklaces[10-6]. However, the first Neanderthals didn't appear until around 430,000 years ago.

2 million years ago
Non-natural artifact: a seashell with a human face crudely carved on it was found embedded in rock in Suffolk, England in 1881. Again, the modern human civilization that created it did not survive.

2 million years ago
Non-natural artifacts: stone tools and other artifacts dating from this era were found during an archaeological dig in Hueyatlaco in Mexico in the 1960s. Their discovery was highly controversial: Virginia Steen-McIntyre, a graduate student who worked at the site, lost her job after revealing them, and she was shunned by the archaeological community afterward. The official report on the archaeological investigation of the site was never published, and the area where the artifacts were found was closed off[10-7].

1.2 million years ago
The first of three modern near-extinctions. The worldwide hominin population collapsed. This may have been caused by over-breeding, or the population might not have been very large in the first place. The total global population was about 26,000, with a breeding population of about 18,000. We would consider a population of this size "endangered." There were numerous small, isolated groups all over the world. The more remote ones eventually died out, leaving only the core groups in Africa.

> This pattern would repeat itself when the main groups of modern humans arrived on Earth.

430,000 years ago
The first-known Neanderthals appeared in Spain (disputed).

400,000 – 500,000 years ago
Missing link. If modern humans had evolved on Earth, the species that preceded us should appear in the timeline at this point. We've found no evidence that such a species ever existed.

400,000 years ago
Modern human DNA from this period has been discovered in northern Spain. There's also evidence that modern humans were living in Israel, and Aboriginal people were living in Australia.

> These must have been the first places on Earth where modern humans were dropped off by the aliens and were able to survive.

350,000 – 300,000 years ago
Modern humans were living in Morocco, and possibly throughout Africa. This group was probably the second main group to be dropped off on Earth. They were well-established by 300,000 years ago, so they must have been brought here long before this – perhaps around 350,000 years ago.

270,000 years ago
An early exodus of modern humans from Africa. They traveled to Europe and appear to have interbred with the Neanderthals. They might also have waged war against them[10-8].

250,000 years ago
According to mainstream scientists, the Neanderthals first appeared. This would be bizarre if it were true, because modern humans had already been here for at least 150,000 years. In fact, if the Neanderthals *genuinely* didn't appear until this point, they might even have descended from *us*.

> The Neanderthals definitely did *not* descend from modern humans. Nor did modern humans descend from the Neanderthals.

250,000 years ago
More stone tools dating from this period and attributed to modern humans were found in Hueyatlaco in Mexico during the 1960s archaeological digs. (See also: 2 million years ago.) Mainstream anthropologists say modern humans didn't reach Mexico until 12,000 to 15,000 years ago. This appears to be the third group that was dropped off by the aliens, but the group doesn't appear to have survived for very long.

208,000 years ago
Y-chromosome Adam appeared. This is an important date in the official human timeline. Many mainstream anthropologists say this person was the ancestor of all modern humans. Others argue that modern humans didn't appear for another 50,000 years. Many non-mainstream anthropologists believe we had already been here for around 200,000 years.

195,000 years ago
Modern humans have been confirmed to be living throughout Africa, and they began migrating across Asia and Europe. (Disputed by mainstream anthropologists.)

The reason for the migration is probably because they had finally learned how to conquer the Neanderthals, and they were expanding into their territory.

The remains of one of these groups was found in Ethiopia in 1967, and they were initially thought to be the first-ever modern humans. But they weren't. Older modern human remains and artifacts had been found all over Africa during the preceding decades. They had been incorrectly recorded as belonging to other species because mainstream anthropologists refused to believe we could have evolved that early.

This appears to have been the fourth – and largest – group of modern humans that the aliens brought to Earth.

170,000 years ago
Mitochondrial Eve, said to be the direct ancestor of everyone alive today, may have lived in Africa.

150,000 years ago
The second of three modern near-extinctions. This one was caused by glaciation. The worldwide human population may have fallen to just 600 people, so we were critically endangered. The main group of survivors lived in South Africa.

140,000 years ago
The first evidence of long-distance trade.

131,000 years ago
Possible evidence of modern humans living in what is now San Diego, California, USA. Researchers have no idea who

these people were or how they got there. They may have been dropped off there by the aliens, but it's more likely they were the last survivors of the Mexican group. They probably died out soon after this.

130,000 years ago
The first-known sea crossing: modern humans reached Crete. There's also evidence of a second exodus from Africa. This group traveled along the Arabian coast[10-9] into India, eventually reaching Australia where there was already a small Aboriginal population.

120,000 – 130,000 years ago
According to mainstream anthropologists, modern humans evolved in Africa. Most anthropologists now accept that this date is wrong, and that modern humans were living in Morocco 300,000 years ago. There was a definite influx of new blood around this time though, which probably marked the arrival of the fifth main group from Eden. I suspect they were brought here to boost our numbers, as we may have been close to extinction.

125,000 years ago
Another early migration from Africa is thought to have stalled around Israel. Tools from this era have also been found in Arabian archaeological digs.

80,000 – 120,000 years ago
Modern humans were living in caves in China. This predates the largest exodus out of Africa by 50,000 to 60,000 years, and suggests that the aliens may have dropped off a sixth group there.

The aliens seem to have been experimenting with new habitats for us, as most of the other groups had died out.

100,000 years ago
Evidence of modern humans living on the Serra da Capivara plateau in Brazil. (Disputed: mainstream anthropologists say we didn't reach South America until 12,000 to 15,000 years ago.) This could have been the Mexican group migrating south, but the worldwide population was dangerously low at this point, and the Mexican group had probably died out. It's more likely that this was the seventh group to be dropped off by the aliens.

100,000 years ago
Non-natural artifact: the Williams Enigmalith (a modern-looking electrical plug embedded in a granite pebble). It was discovered at an undisclosed location in North America in the 1990s. The plug has never been found in any electrical catalog.

72,000 years ago
Clothing and jewelry were invented.

70,000 years ago
The third of the three modern near-extinction events: the Toba super-eruption in Sumatra, Indonesia. This caused six years of global winter. The Earth's human population collapsed again, with no more than 10,000 people surviving, and possibly no more than 1,000.

The aliens had now dropped off at least seven large groups of modern humans on Earth, and hardly any of their members had survived. We were critically endangered and clinging on by a thread. There's no evidence that the aliens brought any further large groups of us here after this, so they may have finally given up trying.

Fortunately, we somehow staged a remarkable recovery.

65,000 – 75,000 years ago
The famous "Out of Africa" exodus that mainstream historians say happened around 60,000 years ago. It almost certainly happened 10,000 years earlier than that when the remaining African groups went searching for new food sources following the Toba super-eruption. An unknown number of people left Africa, traveled north to the Mediterranean, and then turned east and went to Asia. They took over the native hominins' territory and replaced them.

There were a few other small human populations around the world at this time, but they would have been in a desperate and precarious state.

65,000 years ago
A smaller migration from Africa to the Middle East.

50,000 – 60,000 years ago
Modern humans were living in Laos in Indonesia.

55,000 years ago
According to mainstream anthropologists, the first Aborigines colonized Australia. However, DNA evidence indicates that the Aborigines had already been living in Australia for 350,000 years. Few of them would have survived the three near-extinction events though, and they would have

been reduced to a few isolated communities. This date probably marks the point when their population staged a comeback.

50,000 years ago
The Great Leap Forward. This was a cultural revolution in which we began the ritualized burying of the dead, started making clothes, and developed complex hunting techniques.

10,000 – 50,000 years ago
A significant number of Asians returned to Africa. The reason for this is unknown, but we've tracked their movements through their DNA.

40,000 – 50,000 years ago
The first evidence of modern human behavior and cognition: abstract thinking, deep planning, art, ornamentation, music, and blade technology. The "blocks" the aliens placed on our brains when they brought us here appear to have begun wearing off. Or perhaps they were removed by a different race of aliens.

43,000 – 45,000 years ago
Early European modern humans (the Cro-Magnons) appeared. They were genetically identical to us, but more robust. They would later become the first modern humans to develop blue eyes.

40,000 years ago
The Neanderthals became extinct. There was a significant migration of modern humans from Africa to Europe as we expanded into their former territory and joined forces with the Cro-Magnons.

35,000 years ago
The first known cave art was created.

12,000 – 15,000 years ago
The world's first city was established at Tiwanaku, Bolivia.

11,500 – 12,000 years ago
The Mesopotamian civilization was established.
Göbekli Tepe was established in Turkey. Beer was invented.

10,000 – 12,000 years ago
Agriculture began.

10,000 years ago
The first fixed communities (villages) were established. The domestication of dogs began.

7,700 years ago
People with white skin and blue eyes first appeared in northern Europe.

6,000 years ago
A massive leap forward in knowledge: writing, mathematics, architecture, pyramid-building, science, astronomy, modern societies, money, elections, sanitation, and more – including the invention of wheeled vehicles. These all apparently sprang out of nowhere, in multiple places around the world at the same time. We'll look at this in more detail in the next book in the series.

6,000 years ago
The first modern civilization appeared in Sumeria, Mesopotamia (now Iraq and the surrounding area).

5,100 years ago
The Ancient Egyptian civilization was established.

2,200 years ago
The Antikythera mechanism (the first-known analog computer) was made. It was recovered from a shipwreck in Greece in 1902, and incorporates technologies that were unknown until 900 years after it was made.

2,000 years ago
The Roman Empire was established.

600 years ago
The Aztec and Inca Empires were established.

500 years ago
The modern era of human civilization began.

250 years ago
The industrial revolution began.

60 years ago
The space age began.

Our evolution on Earth (since we arrived)

Although we didn't originate on Earth, we've evolved in several significant ways since we arrived[10-10][10-11].

Height

One of the main changes has been in our body size. Most people believe that earlier humans were much shorter than us, and that we only became tall in the modern era. In fact, we were *much taller* when we first arrived on Earth.

This fits with us having come from a planet with a lower level of gravity. As we adapted to the stronger gravity on Earth, we became shorter and more compact. We were at our shortest around 10,000 years ago. And then, once we had finally begun to adapt to our new environment, and we had developed better nutrition and health care, and better standards of living, we began to grow taller again.

Some populations are still growing taller, but in most developed nations the rate has tailed off or stabilized.

> Another important factor is that women generally prefer taller men, so the gene for tallness tends to be favored. This is particularly evident in the Netherlands.

A short timeline of human tallness

- 100,000 years ago: modern humans in Africa were tall with long limbs. It's likely that the modern humans who preceded them were similarly tall.

- 40,000 years ago: male Cro-Magnons averaged 6 feet (1.83 meters) tall, with heavy musculature and much greater body strength than most men today.

- 10,000 years ago: the average European male was just 5 feet 4 inches (1.63 meters) tall. This could have been a reaction to the stronger gravity on Earth compared with the level on Eden. But as we'd been on Earth for around 400,000 years by this point, it's more likely due to other factors. One of the most likely causes is malnutrition. This could be linked to our development of (and reliance upon) agriculture, when we were at the mercy of climate change and crop failure. Some biologists also believe that as we domesticated livestock, new diseases may have crossed over from their population into ours.

- 600 years ago: the average European male was an inch taller, measuring 5 feet 5 inches (1.65 meters). Most researchers believe their height was restricted because of their poor diet and poor health.

- Today: the average European male measures 5 feet 9 inches (1.75 meters) tall. This is still well below the average when we first arrived on Earth, but it's a clear sign that we're adapting to the conditions here. Most researchers attribute the gain in height to better diet and health care. Our genome would also have become more diverse as urban populations boomed. Large cities brought together people who had previously lived in isolated communities where inbreeding was rife.

Brain size

It wasn't just our height that changed; the size of our brains changed too – but perhaps not in the way you would expect. Hominin brains had gradually become larger throughout their history, with each species having a slightly larger brain than the one that preceded it. This was especially noticeable over the past two million years. But in another sign that we didn't evolve on Earth, and we don't fit into its timeline, our brains are *smaller* than the most recent hominin species, including the Neanderthals.

Since we arrived on Earth, our brains have shrunk further, with the most noticeable decrease occurring in the past 6,000 years. Our brains are now around nine cubic inches (150 cc) smaller than when we first arrived. Some anthropologists say this fits with our reduction in height. But even though we've started to grow taller again, our brains are continuing to shrink.

That doesn't necessarily mean we're losing our intelligence though. Researchers have noted that our brain matter is becoming denser, and our smaller brains may allow electrical signals to travel faster.

We can follow this trend using the same timeline that we used to track our changing height:

- 100,000 years ago: the average human brain size was 91 cubic inches (1500 cc).

- 10,000 years ago: the average human brain size was 88 cubic inches (1450 cc).

- Today: our average brain size is 82 cubic inches (1350 cc).

Jaws and teeth

Our jaws and teeth have become smaller too. Again, this has been largely in line with our change in body size. But they've become smaller at a faster rate in the last 10,000 years as we've changed our diet and started eating softer foods. Unlike the other differences we've seen, the general reduction in jaw and tooth size has continued from the other hominin species.

This isn't without issues though: our jaws are now so small that many people have no room for their wisdom teeth, and a significant number have no room for their third molars. In many cases, they have to be surgically removed.

But dentists and anatomists have noticed that increasing numbers of people no longer have wisdom teeth and third molars. It's likely that we will eventually lose them altogether.

> Many countries add fluoride to their water to help improve dental health. It's also routinely added to toothpaste worldwide. Fluoride thickens the enamel on our teeth and makes them slightly larger.

Racial diversity

It's possible that when we lived on Eden, we were all the same race. Race doesn't actually mean very much, as we're all the same species. All it tells us is that our ancestors lived in a warm or cool part of the planet. Their skin took on a certain color, texture or thickness in response to the local environmental conditions – mainly the temperature and the level of sunlight and UV radiation they were exposed to.

Modern humans are divided into five races:

- Australoid – Aborigines and Papuans

- Capoid – Bushmen and the non-Bantu indigenous people of South Africa

- Caucasoid – those with white skin

- Mongoloid – Orientals and American Indians

- Negroid – those with black skin

But things aren't as straightforward as that. Caucasoids living in northern Europe have much lighter skin than those living in southern Europe. For example, someone from northern Finland looks completely different from someone from southern Portugal, even though they're classed as the same race. And, naturally, as we're all the same species, the different races can (and do) interbreed, resulting in millions of people who are "mixed race."

Why might we all have been part of the same race on Eden? Simply because we were all exposed to roughly the same environmental conditions and solar radiation. Eden is probably permanently covered in cloud. If the cloud is thicker in the equatorial regions and thinner toward the poles, the level of sunlight and UV radiation will be more or less the same across the whole planet. If that's the case, there will only be one race.

Our "natural" skin color is dark. When the first modern humans arrived on Earth, they all had dark skin and brown eyes, and they were tall and long-limbed. This remained the case for hundreds of thousands of years. But small differences began to emerge, and became amplified, as groups of people were dropped off in different parts of the world and started to adapt to the local conditions.

Most of the differences emerged quite recently. Light skin only appeared in the last 10,000 years, and blue eyes only appeared 7,700 years ago. This would have been around 30,000 years after we migrated to the far north of Europe.

> Interestingly, many Neanderthals had light skin and blue or green eyes. Yet, even though we have some of their genes, we didn't gain these traits from them. Light skin and blue eyes didn't appear in modern humans until more than 30,000 years after the Neanderthals became extinct.

Light skin and blue eyes can be regarded as a genetic anomaly. But people with this coloring multiplied more than most of the other races, and colonized more of the world than most of the others. So the Caucasoids became the dominant race in many regions.

But it's not just our skins that are different. Our bodies and faces have altered too, in response to the different conditions.

- People living in hotter climates developed longer, thinner limbs that allowed them to lose heat more easily. Those in cooler climates developed stockier bodies to help them retain heat.

- Those in hotter climates developed curly hair, which helps their sweat to evaporate and keep them cool. Those in colder climates developed straight hair, which helps to keep them warm.

- People in hot climates developed flatter noses to help moisten the air when they breathe in and retain moisture when they exhale. Those in colder climates developed longer, narrower noses to help warm the air as they breathe in.

- People living in *really* cold places developed layers of fat on their faces to keep them warm. They also have broad, flat faces that reduce their chances of getting frostbite.

- We can see obvious differences in each race's eyes. Some have adapted to minimize glare and snow blindness. Others have adapted to let in more light.

All of these changes evolved *after* we arrived on Earth.

Our racial evolution is continuing as we migrate around the planet and interbreed with people from different races. We might eventually return to being one single race – a mixture of all of the current races.

On the other hand, new races might appear as a particular trait emerges and becomes amplified. This might happen in response to environmental factors, such as climate change.

But even if we don't return to being one single race, it's likely that our skin will, on average, become darker over the next few generations. This is because of an effect known as the shrinking globe. People from multiple cultures will migrate to large cities in other countries and interbreed with people from other countries and races.

A study in 2007 confirmed that the rate of human evolution (measured by the number of changes in our DNA) has accelerated in the last 40,000 years. The parts of our genome that are changing most rapidly are those connected with skin color and disease.

Reproduction

Over the last 2,000 to 3,000 years, female hip size has increased. That means women can give birth more easily to babies with larger heads.

More recently, the significant rise in the number of Cesarean births has also led to an increase in the size of babies' heads.

It seems inevitable that babies will be born with larger heads in the future. However, if the rise in Cesarean births continues and becomes the standard, women might eventually lose the ability to give birth naturally. Their hips might even start to narrow again.

Over the same period, young women have begun menstruating several years earlier, meaning that they're fertile from a younger age. The age at which the menopause occurs has also risen, meaning that women now remain fertile for several years longer than they used to.

This is just as well, because the average male sperm count has fallen significantly. Whether this is an evolutionary trait, a temporary blip caused by environmental factors, or the beginning of our extinction, remains to be seen.

Other changes

In the last 2,000 years or so, Western societies have evolved lactose tolerance while Eastern societies have not.

There have also been important evolutionary changes in our hair color, insulin levels, and body mass index (BMI).

Our future on Earth

Although we're still evolving, most biologists agree that the rate of change has now slowed significantly. Natural section has been largely halted because of medical intervention. People who would have died before they reached adulthood now survive and are able to have children. So diseases, conditions and genetic traits that would have previously disappeared have been retained in the general population.

All sorts of other evolutionary changes are occurring in different sectors of the population.

- Radiologists and X-ray technicians are evolving a greater resistance to radiation. The same thing is likely to be happening in those who are regularly exposed to other forms of radiation.

- People in sub-Saharan Africa are developing a greater natural resistance to Lassa fever and malaria.

- People who live in high-altitude regions are evolving a better tolerance for living with a lower level of oxygen.

- People who dive deep underwater to collect food, such as the Bajau people of Indonesia, have evolved the ability to hold their breath for much longer than other people[10-12].

- There's also evidence that we're evolving a greater tolerance for the high level of salt in the Western diet, as well as associated issues including high blood pressure and high levels of cholesterol.

- And, as people have started living in greater numbers in cities, our brains have evolved the ability to remember more of their names[10-13].

Some researchers believe our bodies will change as we adapt to living in a technological world, where machines do most of the work and we get less exercise. We will almost certainly become heavier and fatter. The parts of our brain related to communication might grow larger, while those related to thinking, processing and physical movement might shrink.

Eventually, it should become possible (and ethically acceptable) for parents to pre-select their children from a range of embryos. They'll pick the one they want to implant into the woman so she becomes pregnant. They might simply choose whether their child will be male or female. Or they might choose one that will grow up to be attractive and intelligent; gifted in music and the arts; have a pleasing temperament; be resistant to certain medical conditions, depression, and criminal inclinations; and have a long life.

If none of the available embryos fit their requirements, they could have one altered (or even created from scratch) to give it the characteristics they desire.

The attractive, gifted, long-lived children will grow up and have children of their own, and they will inherit those traits too. So we'll evolve in a new direction at a faster pace than we've ever known.

> Selecting, manipulating and creating embryos won't be cheap, of course. If only the wealthy can afford it, *they* might evolve into a new species or sub-species, while the rest of the population continues along the "regular" evolutionary path.

As we move further into the future, the changes in our bodies are likely to become technological rather than physical.

- We might eradicate genetic diseases and conditions by editing them out of our genome.

- We might develop the ability to communicate "telepathically" using wireless signals and electronic implants – and perhaps one day without implants.

- We might use similar technology to upgrade our brains to think faster, increase our memory capacity, instantly learn new skills and languages, and so on.

- And we should be able to replace any parts of our bodies with identical or better ones if they become damaged or diseased – or if we just fancy an upgrade.

As we become a space-faring species and migrate to other planets, the rate of human evolution will accelerate again –

in all sorts of directions. We'll look at this in more detail in the next two chapters.

> It's interesting (though perhaps alarming) to note that our current society *could* be encouraging us toward lower intelligence. Those who gain college and university degrees and go on to have successful careers tend to put off having children until later in life. Consequently, they have fewer children than those who leave school without any qualifications and start having children straight away.
>
> However, many researchers argue that they've found no proof that intelligence is genetic. Children born to people without any qualifications may be just as intelligent as those born to highly qualified parents. They may be less likely to attend college or university, or have successful careers, because of their more impoverished backgrounds. But, overall, there's no indication that our average intelligence is decreasing.

In summary, there are three versions of our long-term future, and they could *all* happen:[10-14]

1. We could remain more or less the same as we are now, but with minor evolutionary changes. All of the different races will probably merge into one. But we will still remain the same species.

2. One or more new human species might evolve. The first of these is most likely to evolve on Mars because of the different environment there. As we colonize other planets, new human species could evolve there too. And, as we've just seen, the very wealthy could also split into a separate species, as they can afford to have "designer" children and upgrade their body parts and genomes.

3. Humans might integrate with machines. We might initially become cyborgs – part human and part machine. But, ultimately, our brains might simply run as software inside a computer or robot.

But as for our long-term future on Earth ... well, we might not have one[10-15].

There's a high probability that sometime during the next few million years a large asteroid will hit the planet, throw up millions of tons of rock and dust, block out the sunlight, and create a winter that lasts for years, if not decades. If that happens, crops will fail, leading to mass famines and millions of deaths across the world, even in highly developed countries.

If the winter lasts for very long, *all* plant life on Earth could die, depleting the oxygen level, ending the food chain, and causing the mass extinction of almost all life on Earth.

On the other hand, the years-long winter might not be caused by an asteroid but by the Earth itself – a supervolcano. There are several supervolcanos on the planet, and any of them could erupt within the next few million years and wipe out all life on Earth.

But let's be optimistic and assume that neither of those things will happen. We might still be doomed.

The Sun's luminosity, and the amount of solar radiation it produces, will increase as it moves into the second half of its life. As a result, silicate materials on Earth will weather at a higher rate, leading to a reduction in the level of carbon dioxide in the atmosphere.

In around 600 million years' time, the level will fall to the point where trees no longer have enough carbon dioxide to photosynthesize, and they'll die off. Smaller plants will be more tolerant, but as the level of carbon dioxide continues to fall, they'll eventually suffer the same fate. Again, this will lead to plummeting oxygen levels, the end of the food chain, and the extinction of most life-forms*.

I'm sure our descendants will have migrated to another planet by then – probably Mars, and almost certainly several other planets too. And, like Noah and his Ark, we'll probably take millions of plant and animal specimens, seeds and embryos with us, so they can continue to live on the new worlds too.

> *We have the opposite problem on Earth at the moment. Some of the trees are dying because there's *too much* carbon dioxide[10-16].

But things aren't looking too good for the Earth. In around one billion years' time, the Sun will be ten percent more luminous than it is today. This will cause a runaway greenhouse effect, similar to what happened on Venus. All of the surface water, including the oceans, seas and lakes will evaporate, the water cycle and carbon cycle will end, and the surface will become a barren desert. Any remaining life will become extinct.

In two to three billion years' time, there's a high probability that the Earth's outer core will cool and become viscous. The inner core will no longer be able to rotate freely, and the dynamo effect that generates the magnetic field will cease, just as it has on Mars and Venus. The Earth will then be exposed to the solar wind, cosmic radiation, and solar flares, and the atmosphere will be eroded.

> We've looked at ways of replacing or reinstating Mars's magnetic field. We could use the same technology here on Earth. But we won't be living here by then, and the Earth doesn't have much of a future after that, so I doubt we will bother.

The Earth's surface will continue to heat up as the Sun becomes more luminous. Four billion years from now, it will be hot enough to melt metal.

And then, in around 5.4 billion years' time, the Sun will exhaust the hydrogen supply in its core and start to become a red giant. Over the next two billion years it will expand dramatically in size, absorbing Mercury, Venus and, in around 7.6 billion years' time, perhaps even the Earth too.

But ... there is another possibility.

As we saw earlier, some astrophysicists believe we might eventually develop the ability to move the Earth into a wider orbit[10-17]. They suggest we could accomplish this by using gravity assists (also known as the "slingshot effect") from multiple asteroids. As the Sun becomes more luminous, we could gradually move the Earth away from it, thus keeping the level of solar radiation constant.

The biggest problem is that the astrophysicists believe it will take us about one billion years to develop this technology. As we've just seen, that would be far too late. We really need to develop the technology within the next 250 million years, if not sooner.

Over the next 110 million years, we expect the Sun's luminosity to increase by one percent. That might not be a bad thing though, because the Earth is expected to cool significantly by then. The problem is that the increase in luminosity will continue.

Within 250 million years, global warming could cause a mass extinction event that we might not be able to recover from.

As we've just seen, within 600 million years, there might not be enough carbon dioxide for trees to be able to photosynthesize.

800 million years from now, the levels of carbon dioxide and oxygen may have fallen to the point where more than ninety-nine percent of life on Earth becomes extinct.

If we can't relocate the entire population to Mars, let's hope our descendants can devise a faster way of moving the Earth into a wider orbit. Otherwise we face an uncertain future.

> If we move the Earth, we'll also need to find a way of taking the Moon along with us. Otherwise we'll lose the stabilizing effect it has on the Earth's axial tilt — among other things. This would cause dramatic shifts in the climate.

Going home

Could we ever return to our original home planet, Eden? It's a dream of many, but I doubt it will ever happen.

- We have no idea where Eden is. It might not be in this galaxy.

- It's too far away. We haven't (officially) set foot on the nearest planets – Venus and Mars. Two of our spacecraft (*Voyager 1* and Voyager *2*) took forty years just to leave the solar system. Until we develop new forms of space propulsion, we won't be able to travel to a habitable planet in a reasonable amount of time.

- We've evolved too far and we aren't pure humans any more. We're hybrids of humans and terrestrial hominins (and perhaps also aliens), our genes have been manipulated to enable us to survive on Earth, and we've evolved in numerous ways since we arrived here. If we returned to Eden, we wouldn't fit in, and we might not even be able to survive there. We're Earthlings now, whether we like it or not. (And we'll almost certainly be Martians in the future.)

- Eden's environment may have collapsed, rendering it uninhabitable. It might have been failing (for natural or man-made reasons) before we left it. That might be why we were removed (or rescued) from it. It might no longer be the lush paradise it was when we lived there, but more like Mars (frozen, dry and barren), or Venus (boiling, dry and barren).

> This is unlikely, as a total collapse takes billions of years. But we've already noted that Eden could be a billion years older than the planets in this solar system. If its star is a billion years older than the Sun, it may have become too luminous for life to survive.

- Eden might no longer exist. The Earth might ultimately be consumed by the Sun when it expands into a red giant. That may have happened to Eden already. If that's the case, there will no longer be a planet for us to return to.

> If Eden still exists, remnants of our original civilization might remain there (unless the aliens have removed it). It may have been taken over by nature when we left, and fallen into ruin. If it was a *really* long time ago, everything might have crumbled into dust by now. Alternatively, the dust might have buried and protected the most important structures and settlements.

> On the other hand, even though tens of thousands of us were brought to Earth from Eden, millions of us might still be living there.

> If we managed to find Eden and wanted to live there again, we might need to terraform it. That process could take 100,000 years or more.

> We would also need to ship equipment and supplies there from Earth (or Mars). But Eden might be so far away that this would be impossible.
>
> Terraforming Mars would be a good test though. If we can do that, then we should be able to terraform other planets too. And we should develop faster and larger spacecraft in the future that will enable us to carry thousands of people and millions of tons of cargo.

- Even if we found Eden, and we could live there or make it habitable again, the aliens who brought us to Earth might prevent us from returning to it.

> The aliens may have spent a small (or large) fortune removing us from Eden and cleaning it up afterward. I don't think they'd want us back.

I'm not going to *completely* rule out returning to Eden. We *might* one day find out where it is, find a way of reaching it, and be allowed to go there. But I don't believe that will happen for millions of years. It's my firm belief that it will never happen at all – but I'd love to be proved wrong.

As the Earth starts to become uninhabitable, the aliens themselves might decide to relocate us. They might, just possibly, return us to the planet we originally came from – especially if they've repaired all the problems with it.

Confused aliens

Some researchers have proposed the theory that the aliens who brought us here may have gotten the Earth and Mars mixed up. According to the theory, an initial scouting mission may have identified Mars as a good potential new home for us. But we were then brought here by a different crew, and they may have dropped us off on Earth by mistake.

It's an interesting theory, but one that can be quickly disproved. For example, the aliens didn't just bring us here *once*, they did it several times, making small changes to our genome each time one of the groups failed to survive.

Conclusion

In the end, we should feel fortunate and grateful that the aliens dropped us off on Earth – and that the Earth existed as a Plan B. If they'd dropped us off on Mars, as they may have originally planned, we would all be extinct by now.

You might think it would have been better if they'd left us on Eden. And you might be right. But if Eden's environment had collapsed like Mars's did, remaining there might not have been possible. It's important to remember that Eden is much older than the planets in this solar system. Its outer core may have begun to cool and solidify, just as Mars's did. And that means no magnetic field, no atmosphere, no water ... and no place like home.

In the next chapter we'll take a closer look at our possible future lives on Mars.

11
Our Future On Mars

We've looked at how we might evolve if we remain on Earth, but as we saw in the previous chapter, we can't remain here indefinitely. At the very least, we'll establish new colonies on the Moon, Mars, and perhaps elsewhere. We won't necessarily terraform Mars, but we *will* need to live there, and establish permanent settlements, even if they're underground or in domes. Human evolution will then diverge, with the citizens of Mars evolving in a different direction from those who remain on Earth.

Among other things, they'll begin to adapt to the lower level of gravity, the lower level of solar radiation, and the higher levels of cosmic radiation and UV. But they could also evolve in all sorts of other interesting and surprising ways.

Gravity

The gravity on the surface of Mars is only thirty-eight percent of the gravity on the surface of the Earth. The average person in the USA weighs around 180 pounds (12 stone 12 pounds, or 81 kilograms). On Mars they would weigh just 68 pounds

(4 stone 12 pounds, or 31 kilograms), which is about the same as a ten-year-old girl on Earth.

> A ten-year-old girl on Mars would weigh less than a two-year-old girl on Earth[11-1].

The first effect of living with the lower gravity on Mars is that our skeletons will become lighter, as they won't need to work so hard to support us. Our bones will become longer, meaning that we will grow taller, and they might also become more hollow. In line with this, we'll lose a significant amount of our musculature, as we'll no longer need it to support our heavy skeletons. As a result, we'll become even lighter, and we'll be able to move around faster and more freely.

> This would be fantastic — as long as we didn't return to Earth. With only a lightweight skeleton and weak muscles, we might only be able to move around by crawling on all fours — while moaning about the unbearable pain.

After two generations of losing our bone density, some researchers believe our bones would become stronger again, yet retain most of their lightness.

Giving birth in the lower gravity should be much easier for new arrivals on Mars, but it could become perilous if the women remained there for more than a year. As their bones became weaker and less dense, their pelvises could shatter during childbirth — and that could be fatal. This would be compounded by the fact that women transfer a significant amount of calcium from their bones to their growing babies during pregnancy[11-2]. Because of this, Cesarean births might be the only option for the first few generations of new Martians.

Over time, the women will lose their heavy musculature and gain flexibility and agility. Their bones should also strengthen again, as we saw above, and normal births should become possible once more.

But we don't fully understand how well we would tolerate the lower gravity on Mars. Not only is it significantly lower than on Earth, it's also significantly lower than on our home planet Eden.

Astronauts such as Scott Kelly, who have spent extended periods in microgravity, have experienced numerous health issues, including bone demineralization and muscle loss, digestion and cognitive issues, blood clots, and eyesight deterioration[11-3][11-4]. Would we suffer similar issues on Mars? If so, to what extent?

Although the gravity on Mars is significantly higher than the microgravity on board the International Space Station, some researchers believe we would suffer exactly the same effects; they'd just take longer to occur.

It's important that we fully understand this before the first permanent settlers set off for their new lives on Mars. In 2001, the Mars Society proposed a spacecraft, the *Mars Gravity Biosatellite*, that would examine the effect of Mars's lower gravity on mammals. The spacecraft would be launched into space, where it would spin at around thirty-two revolutions per minute to simulate the level of gravity on Mars. It would carry a cargo of mice, which would be recovered and examined when the craft re-entered the Earth's atmosphere at the end of its five-week mission. The project was well-supported, and scheduled to launch in 2010 or 2011. But it was abandoned in 2009 because of a lack of funding and NASA's "shifting priorities."[11-5]

Light

A fine, sunny day on Mars would resemble a dull, gloomy one on Earth. This would be great for our eyes, which don't handle bright light well. But since we've been on Earth, we've adapted (to some extent) to living on a planet where the light is too bright.

There are two schools of thought about what might happen if we migrated to Mars. The first says we would enjoy living in the dim, relaxing light, and we'd quickly get used to it. Over time, our eyes would grow larger to let more light in – and they would become freakishly huge by today's standards. Our heads would enlarge too, to accommodate our larger eyes.

The second theory says this wouldn't happen because we would find the dimness too frustrating. Instead, we would supplement the natural light with artificial lighting. It wouldn't be so bright that it would dazzle us or damage our health, as it does on Earth. But it would be bright enough that our eyes wouldn't grow any larger than they are now.

> We would need to use artificial lighting if we lived underground – which we undoubtedly will in the early years of Martian settlement. So the issue of the light being too dim only applies after we've terraformed the planet and started living on the surface. If light was too dim for us, and too dim for crops to grow, then we would use artificial lighting on the surface too.

> Evolutionary biologist Scott Solomon believes that if people are forced to spend their lives underground on Mars, they will become near-sighted. Within a few generations, children will be born with no ability to see at a distance. They'll need to wear eyeglasses if they go outside.

Radiation

The constant exposure to radiation is an interesting issue. Even underground, the levels will be significantly higher than they are on Earth. If someone spends their entire life on Mars, they will be exposed to 5,000 times more radiation than those who remain on Earth. The first few generations of migrants to Mars could therefore suffer from a much higher incidence of cancer, and they may have shorter lives.

> By the time we establish permanent settlements on Mars, we should have developed cures for most forms of cancer. In fact, if I were in charge of the Mars settlement program, I would make sure we had the cures first, before allowing any long-term settlements to go ahead.
>
> Short-term research and exploratory missions shouldn't present too much of a problem. But once we start planning longer-term missions and permanent settlements, even if they're underground, we should pump as much money as possible into curing cancer, and declare it a vital part of the Mars program.

As we saw earlier, the higher levels of radiation and ultraviolet light on Mars will affect our evolution too. The Martian citizens will eventually develop dark skin to help protect them from radiation damage – even if they live underground. They might develop thicker skulls too, to help protect their brains from radiation. They might even develop two sets of DNA, giving them a back-up copy in case one set becomes damaged.

> I don't actually believe this would happen. But we *could* develop "redundant" genes that take over certain vital and specific functions if our original genes become too damaged.

If we terraformed Mars and restored its magnetosphere, most of the harmful radiation would be diverted away from the surface. The Martian citizens would no longer need the protection that dark skin provides, and they might develop white skin again. Terraforming would also reduce the amount of UV reaching the surface, so the Martians would need white skin to help them absorb as much of the UV as possible and create enough vitamin D.

> Many people on Earth take vitamin D supplements (including me). The Martian citizens could do that too. But if we used artificial lighting on Mars, and tuned it so that it supplied exactly the right amount of UV, there should be no need for anyone to take supplements.

> If the Martians used this form of artificial lighting from the outset, or if everyone took vitamin D supplements, their skin might remain dark even after terraforming. They might even retain the different races we have now, or gain new ones.

Birth issues

Not only would giving birth on Mars be dangerous for the first few generations of new mothers; their babies would have a hard time too. The higher level of radiation could cause a significant number of mutations in their DNA – thousands of them compared with around sixty for babies born on Earth. As a result, miscarriages will be far more frequent.

But not all of those mutations will be harmful; some of them might be highly beneficial.

The high number of beneficial mutations will speed up evolution on Mars, and future generations will become well-adapted to living there. In fact, a new species of human could evolve there within tens of thousands of years, compared with the hundreds of thousands of years it would take on Earth.

> Once the people of Mars became a new species or sub-species, it wouldn't be a good idea for them to interbreed with people from Earth – even if it were possible. They could set the evolutionary progress back by several generations.

In fact, interbreeding might be banned. Any children that resulted from illegal interplanetary breeding might be sterilized – or even euthanized – to prevent them from having any children themselves.

> If the citizens of Earth and Mars became different species, they would no longer be biologically compatible, so interbreeding wouldn't be possible anyway. However, as we saw earlier, modern humans and Neanderthals are different species, and *they* seem to have managed to interbreed from time to time. (Although they may have had some help from the aliens.)

> The citizens of Mars might want to avoid all contact with people from Earth anyway, as they might carry infectious diseases that aren't present on Mars.
>
> If everyone traveling to Mars was screened and quarantined before making the journey, no infectious diseases should ever reach it. But the screening and quarantining program would have to be rigorous and continued forever. If a single infected visitor avoided screening and quarantine, they could wipe out everyone on the planet.
>
> This is yet another reason why the Martians might not want to consider traveling back to Earth. The diseases they'd encounter might kill them.

Some anthropologists believe modern humans carried diseases that the Neanderthals had no resistance to. This could be another reason why the Neanderthals became extinct.

If the citizens of the Earth and Mars avoided each other completely, some researchers believe it would take just 250 generations (or about 6,000 years) for the Martians to look substantially different from the people of Earth. But they would retain the ability to interbreed (and would therefore be classed as the same species) for around 100,000 years.

This could happen even faster if we used genetic engineering to fix any issues the Martians were having. For example, if we gave Mars a lower level of oxygen than the Earth to reduce the risk of fire, we could modify people's genes to make them better able to tolerate it. Or we could send settlers who currently live in high-altitude locations on Earth, as they would already be adapted to the lower level of oxygen and the lower air pressure.

As people settle on Mars and other planets, they'll evolve in different directions and eventually become different species. But will we still consider them human?[11-6] Would we ever consider them aliens? Perhaps it depends on how much they look like us.

> Some species of alien are reported to be virtually indistinguishable from us.

> In a billion years' time, when the Earth is no longer habitable, humans might be spread across numerous planets and moons. As each population will have evolved in a different way, *none* of them might technically be "human" any more[11-7].

It's important to bear in mind that we don't yet know whether human fertilization and reproduction will work in space.

If a spacecraft carries a crew to colonize a distant planet, it might take decades – or longer – to reach it. The crew members might need to reproduce during the journey so that their children or grandchildren can colonize the planet when they eventually arrive.

If human reproduction doesn't work in zero gravity or microgravity, even if we use things like in vitro fertilization and artificial wombs, then it *should* work if give the craft artificial gravity. But we haven't tested that yet.

> Artificial gravity should also help overcome some of the other issues astronauts experience during long space missions. These issues include loss of bone density, cardiovascular problems, balance issues, and metabolic issues.

> As we've seen, those same issues could occur on Mars, as the gravity there is significantly lower than it is on Earth.

> A radical solution might be to modify the genomes of candidates selected to become the first Martian settlers – and ensure that those modifications are inheritable. The settlers would arrive on Mars already adapted to living there, and their children would be born with the same adaptations.

Acclimatizing

As we've seen, living on Mars will initially be much like living in an Antarctic research station on Earth. The stations will be heated to the same temperature as they are on Earth, and people will rarely go outside. The same thing will apply if we upgrade to living in underground caves and tunnel networks.

If we build domes on the surface, they'll trap solar radiation, their climates will be computer-controlled, and they should be pleasant and warm. Again, the people living in them shouldn't need to acclimatize in any way (apart from adapting to the lower gravity).

Terraforming Mars would allow people to walk on the surface freely. During the terraforming process, we could adjust the density and composition of the atmosphere so that it trapped the right amount of solar radiation. We could also tune the temperature to be whatever we wanted it to be. We could even make it warmer than the Earth if we wanted to.

But, as we saw earlier, one of the best ways of terraforming Mars might be to give its atmosphere a high level of water vapor. That means it will be cloudy, foggy, and it will probably rain most of the time. So everyone might prefer to stay in their domes.

Vision

The lower gravity on Mars means that long-term visitors and migrants to Mars might suffer from a build-up of fluid and pressure in their heads. As a result, they might develop problems with their vision, as the pressure will affect their ability to focus. It could also cause more severe issues such as glaucoma, and their retinas could become permanently damaged, leading to blindness.

Some researchers have suggested that the levels of UV on Mars could cause cataracts. This is unlikely though. Short term visitors to Mars will wear helmets with UV filters in their visors – just like astronauts do on the Moon. If they live underground or in domes, they'll use artificial lighting and UV filters. And if they live on a terraformed planet, the atmosphere will filter out most of the UV.

Even if fluid build-up and cataracts were a problem for the first few generations of settlers, future generations born on Mars should adapt to it and have normal vision, with no greater risk of blindness than those living on Earth.

> Other solutions might include high-definition artificial lenses, retinas, or eyes that will allow people to see even better than they did before.

Political and societal evolution

Thus far, we've only discussed the *physical* changes that would occur in people on Mars. But its citizens will be separated from everything else that's happening on Earth too. They will develop new cultures and political regimes, new languages – or dialects at least, new religions, new sports, and more.

Our short-term and long-term future on Mars[11-8]

For the first several hundred years, the human settlers on Mars will have a bad time. They'll be smaller and weaker than people on Earth, they'll have significantly more health issues, and they'll have neurological disorders, and shorter lifespans.

Rather than counter this with drugs and genetic remedies, we should regard it as a necessary evil – a stage we will have to go through if we are to make evolutionary progress and fully adapt to living on Mars.

Eventually – and it might take thousands of years – the people of Mars will become stronger and more robust. They'll evolve ways of coping with the Martian environment, and their lifespans will increase again.

We've noted that the people of Mars might develop darker skin as a defense against the higher levels of radiation. But some researchers believe they might develop *orange* skin, as it offers even more protection. The orange pigment would come from carotenoids, which give carrots, pumpkins and sweet potatoes their color. These foods would grow well in the Martian soil, and they could become a staple part of the diet.

All of this assumes that people living on Mars would be exposed to higher levels of radiation. But they might not be. If they live in properly shielded habitats and domes, the levels should be no higher than they are on Earth. If that's the case, they would *not* evolve darker skin or orange skin, and their rate of mutation should remain unchanged. They would still evolve in a different direction from the people on Earth, but that would be primarily due to the lower gravity – and not much else.

Looking further into the future, I'm convinced that we *will* fully terraform Mars eventually. Its gravity will remain unchanged afterward, but replacing its magnetosphere and restoring its atmosphere should cause the radiation and UV levels to fall to similar levels as they are on Earth.

Humans will have been living on Mars for hundreds of thousands of years by the time we achieve this. But, as we've seen, as long as their habitats were properly shielded, the higher levels of radiation and UV should not have affected them.

The only effect of terraforming that they should notice is that they'll be able to go outside without needing protective equipment and breathing apparatus.

> But they may need waterproof clothing or an umbrella.

In the next chapter we'll consider our possible human future in the further-flung reaches of space.

12

Our Future In Space

As we've seen, the Earth will become uninhabitable within the next 600 million years or so. We have a deadline. The entire population needs to have relocated to somewhere else by then. If we stay here and remain a single-planet species, we'll become extinct – along with every other species on Earth.

Our 600-million-year deadline assumes the Earth will remain habitable throughout that period. But it might not. We could be wiped out long before then by an asteroid, a supervolcano, a gamma ray burst, nuclear war, environmental collapse, or something else.

Our long-term survival depends on our ability to find, terraform, and colonize another habitable planet[12-1][12-2]. And the sooner the better. But this is where we hit the first of many problems. There are no other habitable planets or moons in this solar system, and none within reach that we could make habitable.

Terraforming Mars is undoubtedly our best option for the next few billion years. Even if fully terraforming it is beyond our abilities, we could at least build massive cities underground, and in domes on the surface, and hang out there for as long as we're able to.

But eventually we'll have to leave Mars too. In about five billion years, the Sun will begin to expand into a red giant star, and Mars will become so hot that its surface will sizzle. We wouldn't be able to survive there even if we lived deep underground.

The good news is that as the Sun grows larger and hotter, the solar system's habitable zone will move outward. As a result, some of the moons of Jupiter, Saturn, Uranus and Neptune might then become habitable[12-3].

The bad news is that it will take a considerable amount of time for these frozen worlds to thaw out, for life to become established there, and for them to become suitable new homes for us. In the meantime, if Mars is no longer habitable, we'll be homeless. In fact, we might be homeless for millions of years, and perhaps even for billions of years. Where the heck will we live?

> We will need to terraform the potentially habitable moons of the outer solar system as soon as conditions allow. We will have had plenty of time to develop better technologies by then, so we should be able to do it much faster and more easily than when we terraformed Mars.

> Few of these moons will actually be habitable[12-4]. Of Jupiter's four largest moons, only Callisto is a realistic candidate. The others are too radioactive, or subject to massive tidal forces. Saturn's moons Enceladus and Titan are more promising candidates, as is Neptune's moon Triton.

The next best place to look for a new home is a habitable planet orbiting another star. This is where we hit the second of the many problems we'll need to solve if we are to survive: space is *really* big. The next nearest star, Proxima Centauri, is just over four light years* away. It is unreachable using any of the space propulsion systems we currently have or are likely to develop in the foreseeable future.

> *Four light years = 23.5 trillion miles (23,500,000,000,000 miles) or 37.8 trillion kilometers (37,800,000,000,000 kilometers).

> *Voyager 1* and *Voyager 2* took forty years to reach the edge of our solar system – a distance of 19.5 light hours**. It would take them 70,000 years to reach Proxima Centauri.
>
> **19.5 light hours = 13 billion miles (13,000,000,000 miles) or 20.9 billion kilometers (20,900,000,000 kilometers).

> There are billions of Sun-like stars in the galaxy. According to NASA[12-5], more than half of them probably have exoplanets in their habitable zones. They won't all be habitable, of course, but a good number of them should be.

Let's compare the distance between the Earth and Proxima Centauri with something we're more familiar with: the USA. Let's imagine that the Earth is New York City and

Proxima Centauri is Los Angeles. The distance between them is 2,790 miles (4,490 kilometers). In forty years of traveling, we've only just covered the first mile.

> You might think that's a snail pace, but a snail can travel a mile in a week. It's actually 2,000 times slower than a snail's pace. We truly have a long way to go.

The highest speed one of our spacecraft has ever achieved (as of this writing) is 157,100 mph[12-6]. (This was achieved by using the Sun as a slingshot.) Even traveling at that speed, it would take 18,000 years to reach Proxima Centauri.

But Proxima Centauri doesn't appear to have any habitable planets orbiting it. We might need to travel forty or more light years to reach a planet we can live on. And it might take us as long as 200,000 years to get there.

The spacecraft that sets off on such a journey would have to be massive, perhaps carrying a million people or more. None of the people on board will live long enough to reach the new planet and colonize it; that will be their distant descendants' job. But 200,000 years is an *awfully* long time. Those distant descendants might not be the same species by the time they arrive. Nor would they speak the same language as when they left the Earth[12-7] – which could make it rather difficult to communicate with them.

> Another issue is that the crew's motivation might change with each passing generation. They might resent the fact that their long-dead ancestors chose to be blasted into space, meaning that they themselves would not only spend their entire lives

> in space, but would die there too. They might mutiny, refuse to procreate (meaning that there would be no succeeding generations to populate the new planet) or do something else to jeopardize the mission.

Fortunately, if we're able to make Mars our new home within the next 600 million years, we should be able to stay there for the next five billion years or so. By the time we need to consider migrating to a planet orbiting another star, we should have developed much faster propulsion systems, as well as bigger and better spacecraft, and more efficient fuels.

Some engineers have calculated that a single gram (0.035 ounces) of antimatter could propel a spacecraft to Proxima Centauri (or the much more promising Alpha Centauri B) in as little as twenty-five years. Physicists have been able to create antimatter since 1955, but it would take 100 million years to create a gram of it using their current technology.

Other engineers calculate that it would take several *gallons* of antimatter to complete the journey, not just a single gram. But with several billion years to develop and refine the process, we should be able to create more than enough of the stuff by the time we really need it.

Naturally, we wouldn't send a spacecraft full of people to another planet without checking out that planet first. So, once we've developed a propulsion system that can take us there in a reasonable amount of time, our next task will be to send unmanned craft to every star within forty light years of Earth.

When they arrive, they'll search for potentially habitable planets and, if they find one, land on it, take readings, and send the data back to Earth. Our scientists will then need to decide whether it would be worth terraforming that planet – and how the heck they would do such a thing when it might be dozens of light years away.

Bear in mind that it might take us a minimum of several hundred years to reach that planet, even with the technology we might have developed by then. If it takes us twenty-five years to reach Alpha Centauri using the antimatter-powered craft we looked at above, it would take us 250 years to reach a planet forty light years away.

What if one of the craft landed on a planet and discovered it already had life? Would we take that as a good sign and go ahead with our plans? Or would we steer clear of it to avoid contaminating it? I guess it would depend on whether any of the other craft found habitable planets. If they only find the one, we might have no option but to take it over.

And if it happens to be occupied by a race of advanced hominins like us, we might have no option but to invade it and fight them for it.

> I'm sure we'll ask them *very nicely* if we can move in alongside them before we launch an attack.

It's highly unlikely that any of the nearest stars will have habitable planets orbiting them. If we manage to find one within forty light years of Earth, and it takes us hundreds of years to reach it, terraforming it might be way beyond our abilities. Even if we found a potentially habitable planet just *four* light years away, terraforming it might be way beyond our abilities.

But, on the other hand, it might not. If we can migrate to Mars before the Earth becomes uninhabitable, we should have five billion years to develop the technologies we'll need.

And we could send sentient robots to terraform the new planet. They might locate nearby asteroids that contain the materials they need, move them into orbit around the planet, mine them, and ferry the materials down to the surface. They'd use the materials to terraform the planet, make it habitable (to humans *and* other life-forms from Earth), build habitats, life-support systems, farms, and everything else we'll need.

Meanwhile, other sentient robots could construct a fleet of massive spacecraft in orbit around Mars – or wherever we happened to be living at the time. Once our new planet was ready, we could hop aboard the spacecraft, sit back, and enjoy the ride to our new home.

Of course, if it takes hundreds of years for each craft to reach the new planet, it would be our descendants who settle there, rather than us. But at least they would still be members of the same species. And there's less chance they would mutiny or try to sabotage the mission.

> We're talking about events that might happen five billion years from now. We will have evolved into an entirely different species by then anyway.

While I was researching this book, I came across many scholarly articles that explained how the people on board a spacecraft traveling to a distant planet would live in zero gravity for such an extended period[12-8]. The articles described how we might live, work, exercise, go bowling, practice arts and crafts, and even carry out surgery. ("One problem was that, during open surgery, the intestines would float around.")[12-9].

While the authors of those articles must have collectively spent thousands of hours thinking about these issues, the simple fact is that everything would work just as it does on Earth now. We *would* have gravity. We saw earlier that we could build large, ring-shaped or cylindrical spacecraft and rotate them at a constant rate to create artificial gravity. If we get the speed of rotation right, everything should feel exactly as it does on Earth.

Well, maybe not *exactly* as it does on Earth, because the level of gravity is too high for us. It would be better to rotate the craft more slowly, so the level of gravity matches our original home planet, Eden. We could also reduce the air pressure inside the craft to match the air pressure on Eden.

These craft would be truly *massive* — far bigger than our largest ocean liners. They might be more like flying cities. Traveling on them would be a pleasure, and the people on board them might forget they're even on a spacecraft.

Such craft are way beyond our current level of technology, of course. We might not be able to build them for tens of thousands of years, or perhaps even longer than that. They would also be too large to launch from Earth or Mars, so they would need to be constructed in space, and passengers would need to be ferried into orbit to board it. But traveling into orbit will be commonplace by then.

For the next several hundred years, we'll probably journey no further than Mars or the potentially habitable moons of Jupiter or Saturn (Callisto and Triton respectively). We'll travel in spacecraft that are no larger than today's largest airliners. And I don't believe they'll rotate — so the people on board them will travel in zero gravity.

Until we get around to building massive rotating spacecraft, we *might* actually need to carry out surgery in zero gravity. Imagine if an astronaut traveling to Mars suffered a ruptured appendix six months into the journey. Bringing him back to Earth for surgery wouldn't be possible. Keyhole surgery carried out by robots would be the preferred option, if there was any choice. The procedure would need to be carried out inside a sealed bubble with its own air supply to keep contaminated blood and other fluids away from the rest of the crew*. In the early years, the robot might have to be operated by a surgeon on Earth. That could be problematic, as the surgeon's commands could take several minutes to reach the robot. Eventually, we will develop robots that can carry out surgery all by themselves.

*Another option would be to attach a small bubble to the patient's abdomen, and insert the instruments through small holes. Human surgeons might struggle to see what they're doing, but robots should be able to cope.

If someone became ill on one of the first missions to Mars, and robot surgeons weren't available, one of the other astronauts would have to carry out the surgery. At least one of the astronauts would need surgical training – and ideally at least two of them, just in case the surgeon himself needed surgery.

There are other ways of transporting large numbers of people to planets orbiting other stars – though, once again, it might take us thousands of years to develop that level of technology.

For example, everyone on board the spacecraft could be placed into suspended animation[12-10]. The computers controlling the mission would look after their life-support systems and wake them up once they reached their new home. They might have been traveling for thousands of years, yet they would not have aged or had any awareness of the passing of time.

Another option would be to send frozen human embryos or DNA to the new planet. They would be accompanied by robotic systems that would gestate, nurture and educate them once they reached their destination. There would be no adults on board the spacecraft.

> This might be how the aliens transported the first modern humans to Earth hundreds of thousands of years ago.

In both cases, the new planets would have been terraformed and made habitable by sentient robots before the new inhabitants arrived.

The big question we need to consider is: how far can we actually get (in terms of distance) in the next five billion years? Will we reach the nearest stars? Or will we venture further?

What if the unmanned spacecraft determine that there are no habitable planets that we could reach even if we spent thousands of years traveling to them? There would be no point in sending anyone to them.

Will we manage to create wormholes that allow us to travel from one side of the universe to the other in a fraction of the time it would normally take?

Wormholes are great in theory, and we might one day succeed in sending electronic data through them, allowing for near-instant communications. But I can't ever foresee people traveling through them. In any case, who would be at the other end to welcome them? What if there were no planets there? Would the brave souls who volunteered to travel through the wormhole ever be able to get back again? If they couldn't, would we ever know what became of them?

Many people believe we will never be able to leave our current solar system. Remember the prison planet theory? Whoever brought us here might have no intention of letting us leave. We've seen that there's a wall of superheated plasma surrounding the solar system. NASA's spacecraft passed through it unscathed, but *we* might not be able to. If we can't, we could be trapped here[12-11]. And if that's the case, then we are, ultimately, doomed.

Earlier, I posed the question: what might become of us if we were living on Mars when the Sun became a red giant, and it rendered Mars uninhabitable? We might *eventually* be able to live on one (or more) of the moons around Jupiter, Saturn, Uranus or Neptune, once they'd thawed out, but that could take millions of years. Our only other option would be to build super-massive artificial worlds in space.

A good example of a practical artificial world is an O'Neill cylinder[12-12][12-13].

> O'Neill cylinders were conceived by the American physicist Gerard K. O'Neill in his 1976 book *The High Frontier: Human Colonies in Space*. The British science fiction writer Arthur C. Clarke came up with something similar in his 1973 novel *Rendezvous with Rama* – although the ones in his story were created by aliens rather than by humans.

O'Neill cylinders consist of a pair of linked, counter-rotating cylinders, each twenty miles long and five miles in diameter. The rotation generates artificial gravity on their inner surfaces. The counter-rotation cancels out any gyroscopic effects, so that each pair can be steered easily and kept pointed at the Sun so it can receive ample solar energy.

Each cylinder is climate-controlled and can house up to a million people. Dr. O'Neill proposed that the climate inside each cylinder should be like that of Maui, Hawaii on its best day of the year.

Inside each cylinder there could be cities, farms, beaches and even mountains. But of course there would be no natural disasters: no floods, no hurricanes, no earthquakes, no volcanoes.

Jeff Bezos, the founder of Amazon and the space technology company Blue Origin, said: "People are going to want to live there. But if necessary, they could fly back to Earth."

O'Neill cylinders would certainly make fantastic vacation destinations, as well as permanent residences for the rich and famous. But, ultimately, flying back to Earth would *not* be an option; everyone might need to live in them permanently. And if that were the case, we would need thousands of them.

> Each pair of cylinders could be joined to dozens of others to create massive interconnected networks or artificial countries.

Of course, there might not be enough raw materials on Earth (or Mars) to build that many cylinders. The (robotic) builders and engineers working on the cylinders might need to mine other planets, moons and asteroids for materials, and even build the cylinders there.

Another interesting idea, which might not be technically classed as human space exploration, would be to seed the universe with our DNA[12-14].

As we saw earlier, life on Earth (or Mars) may have begun when life-bearing rocks were ejected into space from another planet and landed here. Could we do the same thing ourselves?

Some researchers have suggested that we could implant our genome into an extremophile organism that can survive in space. We would also need to implant the instructions to assemble that genome into a fully functional human when it reached its destination. We would place a colony of the organisms in a fissure in a rock, and fire it into space – along with millions of other similar rocks. One of those rocks *might* one day land on a habitable planet and populate it with humans.

Or, more likely, not.

It's a fascinating thought experiment, but I can't see how it could have any practical application. The genetic instructions contained inside the organism might allow it to create a human embryo, but how would that embryo grow into a fetus without a womb to protect and nourish it? How would the baby survive in the unlikely event that it was born? Even if thousands of embryos reached a habitable planet, and they somehow grew into children, they would have no parents or guardians to feed or care for them. They wouldn't survive, and they certainly wouldn't reproduce and populate the planet.

A better idea would be to send millions of small, sentient robots into space. Each robot would carry heavily shielded phials of frozen embryos that it would defrost and nurture if it ever landed on a habitable planet[12-15].

> The robots would be deactivated when they were launched into space. They would only reactivate if they landed on a suitable planet.

However, in my opinion, the only practical way of accomplishing something like this would be to step back into prehistory. We could send millions of life-bearing rocks into space, but they wouldn't carry *our* genomes, they would carry the genomes of more primitive organisms that can survive on their own.

They might be anaerobic microbes, extremophile bacteria, or some other kind of microorganism. They'd seed the new planet with life from Earth, and eventually, perhaps several billion years later, something like modern humans might evolve from them.

In the final chapter, we'll look at our ultimate future and consider what might become of us when the stars go dark.

13
Our Ultimate Future And Fate

So, we come to the ultimate question: how long can we actually survive?

First, let's assume we won't get wiped out by an asteroid strike, supervolcano, gamma ray burst, environmental collapse, nuclear war, or alien invasion.

As we've seen, if we remain on Earth, we have a maximum of about 600 million years left. After that, the Sun's increasing luminosity will cause the level of carbon dioxide to fall too low for trees to be able photosynthesize. Other plants will suffer the same fate over the next few million years. And that will be the end of the oxygen supply, the food chain, and all higher forms of life.

Some astrophysicists believe we could one day move the Earth into a wider orbit. Unfortunately, they estimate it will take about a billion years to develop the technology. It's unlikely we'll survive that long.

And developing the technology is one thing, but how long would it then take to move the Earth into a safer orbit? To stay *completely* safe, we'll need to develop the technology much faster than the astrophysicists estimate. Ideally, we'll need to begin edging the Earth into a wider orbit within the next 250 million years.

But there are other options. We could live on Mars – regardless of whether or not we've managed to terraform it. Most estimates suggest that terraforming it will take about 100,000 years. By that time, we should also have developed spacecraft large enough and fast enough to transport vast numbers of people there. We should also have developed ways of building habitats on Mars – either on the surface, or in enormous domes or caves. We'll be a two-planet species – for as long as the Earth remains habitable. After that, we should be able to remain on Mars until the Sun begins to expand into a red giant in around 5.4 billion years' time.

> The Sun will become more luminous throughout this period, so we will need to adjust Mars's atmosphere in response. But we should have the technology to do this: it will be an offshoot of the technology we use to terraform Mars.

> If terraforming fails, we'll just need to make minor adjustments to the climate inside our domes and other habitats.

When the Sun begins to expand into a red giant, Mars will also become hot and unhabitable. No amount of terraforming and climate adjustment can save us from that. If we are to survive as a species, we'll have to move again.

At this point, the outer planets of the solar system will start to fall into the Sun's habitable zone. We obviously can't live on Jupiter, Saturn, Uranus or Neptune because they're gas giants, but we *might* be able to live on some of their moons. We just need to hope they become warm enough, and that we can terraform them (or build giant habitats on them), before Mars becomes too hot. If not, that could be the end of us.

But probably not, because we may be able to move everyone into giant, space-based colonies – such as the O'Neill cylinders we looked at earlier.

But *that* isn't a permanent solution either. The Sun won't remain a red giant for very long[13-1]. Red giants burn through the helium in their cores really quickly. We might have as little as a few thousand years or as long as a billion years, but once the helium is used up, nuclear fusion ceases.

At that point, the Sun will shrink until a new helium shell reaches its core. The helium will ignite and the Sun's outer layers of gas and dust will be blown off to form a planetary nebula. As it continues to shrink, it will become a white dwarf[13-2]. But the solar radiation it emits will be too weak to sustain life on any of the remaining planets and moons – or the O'Neill cylinders.

Unless we've managed to reach a habitable planet outside the solar system by then, we will cease to exist. That means we have a maximum lifespan of about nine billion years.

But let's be optimistic and assume we *can* find another habitable planet outside the solar system. We'll also need to find a way of terraforming the new planet, getting everyone through the wall of plasma surrounding the solar system, and transporting everyone to their new home.

13. Our Ultimate Future And Fate

Our best bet for our long-term survival would be to find an Eden-like planet orbiting a red dwarf star. Red dwarfs burn their fuel really slowly, and they can last for 100 billion years or more[13-3]. The good news is that about eighty-five percent of the stars in our galaxy are red dwarfs. Even better, astronomers believe there could be tens of billions of super-Earth planets in their habitable zones[13-4].

> There should be a similar number of super-Mars planets too. Those are the ones we should *really* be focusing on.

There are six red dwarf stars within ten light years of Earth:

1. Proxima Centauri (4.2 light years from Earth)
   The nearest star to Earth after the Sun

2. Barnard's Star (5.95 light years from Earth)

3. Wolf 359 (7.86 light years from Earth)

4. Lalande 21185 (8.3 light years from Earth)

5. Luyten 726-8 (8.7 light years from Earth)

6. Ross 154 (9.68 light years from Earth)

If one of these stars turns out to have a habitable planet orbiting it, and we can reach it and terraform it in the time we have left (or build massive habitats there), then we could, potentially, survive as a species for 100 billion years or more.

But could we survive beyond that? Maybe.

New stars will continue to be born for the next 100 trillion years or so — though at a much slower rate than they are today[13-5]. We might be able to hop to planets orbiting newer stars as the older ones decline.

In about 120 trillion years' time, though, our options will become severely limited. Most of the stars will have gone out, and only a hundred of them might be visible in the night sky. They too will disappear over the next trillion years or so*. And then the entire universe will be completely dark.

> *Assuming some of them are M-dwarfs, which have trillion-year lifespans.

> We might run into another problem here though. One researcher has calculated that Earth-sized planets orbiting M-dwarf stars might suffer from intense hurricanes[13-6]. Hopefully, a Mars-like planet (smaller in size and orbiting a little further away from the star), would fare better.

Could we survive beyond that? Again, maybe.

Although the universe will be dark, it will contain plenty of material we could use to create energy. The planets, moons, asteroids, and remnants of the former stars will still be there. If we can find them (in the dark) we could mine them, convert the materials into energy, and keep our planet or space community alive for trillions of trillions of years.

> One of the benefits of living in a dark universe is that we won't have to worry about things like solar wind and cosmic radiation. There won't be any. If we happen to be living on a planet without a magnetic field, that won't be a problem either.

We won't be able to survive forever though, because the universe (probably) won't last that long. Or, at least, we don't think it will.

In fact, we have no idea what the universe's ultimate fate will be. It probably won't collapse back into a single point, as some astronomers have theorized. But there's a very small possibility that it will.

The most likely timeline is roughly as follows:

1 quadrillion (10^{15}) years
Solar systems will cease to exist. The Sun will be a black dwarf.

10 billion billion (10^{19}) years
Galaxies will no longer exist. All stars will be black, and they'll be flung out of their orbits or consumed by black holes.

10^{40} years
Nucleons will decay and only photons and leptons (including electrons and neutrinos) will remain, although they'll be few and far between.

10^{100} years
The last black holes will evaporate.

Realistically, somewhere around 10^{30} years will be our ultimate expiry date. There's no way we'll be able to survive once the former stars, planets, moons and asteroids have disintegrated and decayed.

... Unless we can somehow find another universe and cross over into it.

Some theorists believe the universe won't remain empty forever though. They've calculated that in $10^{10^{10^{56}}}$ years there could be a spontaneous inflation event (another "Big Bang") in which a new universe will be created[13-7]. Or perhaps an infinite number of universes will be created.

Billions of years later, humans might evolve all over again.

> If there are an infinite number of universes, the Earth will form again in at least one of them – and so will you and I.
>
> So, I look forward to seeing you again in $10^{10^{10^{56}}}$ + 13.8 billion years.

14
References

Dead links? No problem!

All of the websites listed in this section were accessed shortly before going to press and were confirmed to be working. If you find that a link no longer works, you should still be able to view an archived copy. Just copy the link, go to www.archive.org and paste it in there. If there's a choice of dates for when the item was indexed, I recommend selecting the earliest one.

In the unlikely event that there's no archived copy, other sites will usually have the same information. Use your favorite search engine to look for the main keywords from the page's title (for example: oldest human fossil).

Credibility of the sources

Where possible, I've only included links to sources that are regarded as credible, or where the information is available nowhere else. Some readers might consider some of the links less than credible at first glance (for example, Wikipedia), but they're included because they amalgamate or simplify complex information, provide useful images, and cite links to credible sources. I excluded sources that didn't cite other credible sources or weren't themselves an original, credible source – apart from a handful of exceptions where not citing a source would have been seen as a serious omission.

The accuracy of the information in this book is very much dependent on the accuracy and reliability of these sources. In cases where the information isn't available elsewhere, and couldn't be verified independently, I've taken it entirely on trust and used it in good faith.

Chapter 1

[1-01] IFLScience (23 August 2019) Tom Hale
There are planets with more life than Earth, research suggests
bit.ly/supermars-1-01

[1-02] New Scientist (5 May 2020) Donna Lu
An ancient river on Mars may have flowed for 100,000 years
bit.ly/supermars-1-02

[1-03] Mysterious Universe (7 August 2020) Jocelyne LeBlanc
Mars may have been covered in ice, not flowing rivers
bit.ly/supermars-1-03

[1-04] New Scientist (21 November 2012) Victoria Jaggard and Joanna Carver
Mars is safe from radiation – but the trip there isn't
bit.ly/supermars-1-04

[1-05] Phys.org (21 November 2016) Matt Williams
How bad is the radiation on Mars?
bit.ly/supermars-1-05

[1-06] Futurism (12 May 2020) Dan Robitzski
Scientists say they've found the perfect spot for a Mars colony
bit.ly/supermars-1-06

[1-07] Futurism (6 August 2020) Dan Robitzski
Scientists: Martian lava tubes large enough to fit planetary base
bit.ly/supermars-1-07

[1-08] New Scientist (24 July 2020) Alice Klein
Mould from Chernobyl nuclear reactor tested as radiation shield on ISS
bit.ly/supermars-1-08

[1-09] Futurism (30 September 2019) Victor Tangermann
NASA's Chief Scientist is oddly terrified by finding life on Mars
bit.ly/supermars-1-09

[1-10] Futurism (2 November 2018) Kristin Houser
Mars used to be dotted with life-friendly lakes
bit.ly/supermars-1-10

[1-11] Gizmodo (23 September 2019)
Ryan F. Mandelbaum
Magnetic field on Mars mysteriously pulses at night, NASA's InSight lander finds
bit.ly/supermars-1-11

[1-12] Wikipedia (accessed 20 November 2020)
Geological history of oxygen
bit.ly/supermars-1-12

Chapter 2

[2-01] Mysterious Universe (1 July 2019) Jocelyne LeBlanc
Life could have existed on Mars before it did on Earth
bit.ly/supermars-2-01

[2-02] Futurism (21 February 2020) Jon Christian
Bill Nye: Humans may be descendants of ancient Martians
bit.ly/supermars-2-02

[2-03] Futurism (25 June 2019) Kristin Houser
Mars had a chance to grow life 4.4 billion years ago, study shows
bit.ly/supermars-2-03

[2-04] Daily Mail (5 November 2020) Stacy Liberatore
Is there life on Mars? Dozens of microbial species discovered 11 inches below the surface of Earth's most arid desert may suggest there are organisms hiding in the Martian planet
bit.ly/supermars-2-04

[2-05] Science Alert (22 November 2019) Nick Longrich
Nine species of human once walked Earth. Now there's just one. Did we kill the rest?
bit.ly/supermars-2-05

[2-06] IFLScience (23 September 2019) Madison Dapcevich
Venus may have been habitable to life billions of years ago
bit.ly/supermars-2-06

[2-07] Mysterious Universe (26 September 2019)
Jocelyne LeBlanc
Venus may have had oceans and Earth-like climate for billions of years
bit.ly/supermars-2-07a

[2-08] Space.com (23 September 2019)
Samantha Mathewson
Venus may have supported life billions of years ago
bit.ly/supermars-2-08

[2-09] University at Albany (accessed 23 November 2020)
Ozone
bit.ly/supermars-2-09

[2-10] NASA (5 March 2015) Release 15-032
NASA research suggests Mars once had more water than Earth's Arctic Ocean
go.nasa.gov/3nKUhxp

[2-11] Gizmodo (15 March 2019) George Dvorsky
In 1997, NASA's Pathfinder mission unknowingly landed near the shores of an ancient Martian sea
bit.ly/supermars-2-11

[2-12] Medium (22 October 2019) James Maynard
The Great Salt Lakes of Mars
bit.ly/supermars-2-12

[2-13] New Scientist (27 March 2019) Yvaine Ye
Mars used to have massive flowing rivers twice as wide as Earth's
bit.ly/supermars-2-13

[2-14] New Scientist (4 January 2017) Lisa Grossman
Mars should have loads more water – so where has it all gone?
bit.ly/supermars-2-14

[2-15] New Scientist (16 July 2008)
Asteroid switched Mars's magnetic field on and off
bit.ly/supermars-2-15

[2-16] Wired (20 January 2011) Lisa Grossman
Multiple asteroid strikes may have killed Mars's magnetic field
bit.ly/supermars-2-16

[2-17] Wikipedia (accessed 23 November 2020)
Geomagnetic reversal
bit.ly/supermars-2-17

[2-18] Smithsonian Magazine (7 August 2019) Emily Toomey
Earth's magnetic field could take longer to flip than previously thought
bit.ly/supermars-2-18

Chapter 3

[3-01] Futurism (28 July 2020)
The danger of blue light is real. Protect your eyes with blue light glasses
bit.ly/supermars-3-01

[3-02] YouTube (23 August 2012) RevZone
Dyskinesia and Blue Lenses
(Note: some people may find this video distressing)
youtu.be/h3y13erwfxw

[3-03] StackExchange Biology (21 April 2014)
What is the minimum air pressure the human body can tolerate if oxygen supply is not an issue?
bit.ly/supermars-3-03

[3-04] Instituto Pio XII (accessed 23 November 2020)
bit.ly/supermars-3-04

[3-05] Mercy Medical Center (1 July 2018)
42% Percent of Americans Are Vitamin D Deficient. Are You Among Them?
bit.ly/supermars-3-05

[3-06] Quora (16 March 2016)
various comments on message board
Are humans the only large mammals that mate year round?
bit.ly/supermars-3-06

[3-07] Daily Mirror (14 July 2018) Grace Macaskill
Hay fever warning issued for pets as pollen hits all time high – these are the types worst affected
bit.ly/supermars-3-07

[3-08] Nature (2013) Briana Pobiner
Evidence for meat-eating by early humans
go.nature.com/35YiUAo

[3-09] That Thinking Feeling (29 April 2019) Clare Jonas
Why do we have food preferences?
bit.ly/supermars-3-09

[3-10] Mysterious Universe (21 April 2020) Paul Seaburn
If space travel causes brains to expand, humans may need alien skulls
bit.ly/supermars-3-10

[3-11] Lifehacker (8 August 2020)
Meghan Moravcik Walbert
Why you should lie (to yourself) about your age
bit.ly/supermars-3-11

[3-12] Mysterious Universe (11 July 2020) Paul Seaburn
Science proves farm animals really can sense earthquakes before they happen
bit.ly/supermars-3-12

[3-13] Gizmodo (1 October 2019) Ryan F. Mandelbaum
Now you can listen to marsquakes
bit.ly/supermars-3-13

[3-14] Futurism (26 June 2019) Victor Tangermann
New theory: marsquakes could be caused by frozen groundwater
bit.ly/supermars-3-14

[3-15] IFLScience (30 December 2019) Alfredo Carpineti
Scientists find source of Mars' strongest quakes
bit.ly/supermars-3-15

[3-16] Gizmodo (25 February 2020) Ryan F. Mandelbaum
There's a lot more happening inside Mars than we knew
bit.ly/supermars-3-16

[3-17] New Scientist (12 November 2020) Leah Crane
Dust storms on Mars are tossing water from its atmosphere into space
bit.ly/supermars-3-17

[3-18] University of Illinois at Urbana-Champaign (accessed 23 November 2020)
Thorsten Ritz and Klaus Schulten
The magnetic sense of animals
bit.ly/supermars-3-18

[3-19] Nanyang Technological University, Singapore (13 October 2018) Shihui Chong
Research: my children don't eat vegetables: why? Consequences & smart solutions
bit.ly/supermars-3-19

[3-20] BBC Good Food (accessed 23 November 2020)
Frankie Phillips
Sugar addiction and children
bit.ly/supermars-3-20

[3-21] Wikipedia (accessed 23 November 2020)
Colonization of Mars
bit.ly/supermars-3-21

[3-22] Mysterious Universe (22 June 2020) Paul Seaburn
A spaceship with just 110 people could colonize Mars
bit.ly/supermars-3-22

[3-23] Futurism (30 June 2020) Dan Robitzski
According to new equations, a Mars colony would need this many people
bit.ly/supermars-3-23

Chapter 4

[4-01] Vox (19 February 2015) Dylan Matthews
These are the 12 things most likely to destroy the world
bit.ly/supermars-4-01

[4-02] Earth System Research Laboratories (accessed 23 November 2020)
Teacher background: natural climate change
bit.ly/supermars-4-02

[4-03] Live Science (25 March 2017) Laura Geggel
How often do ice ages happen?
bit.ly/supermars-4-03

[4-04] Universe Today (9 May 2016) Matt Williams
Will Earth survive when the Sun becomes a red giant?
bit.ly/supermars-4-04

[4-05] MPH Online (accessed 23 November 2020)
Outbreak: 10 of the worst pandemics in history
bit.ly/supermars-4-05

[4-06] Wikipedia (accessed 23 November 2020)
Estimates of historical world population
bit.ly/supermars-4-06

[4-07] The New York Times (14 May 2020)
Ralph Blumenthal and Leslie Kean
Navy reports describe encounters with unexplained flying objects
nyti.ms/35Vary0

[4-07a] BBC (23 December 2020) John Sudworth
Covid: Wuhan scientist would 'welcome' visit probing lab leak theory
bbc.in/34Gx9Jj

[4-08] NBC News (6 May 2019) Denise Chow
1 million species under threat of extinction because of humans, biodiversity report finds
nbcnews.to/33cjZTv

[4-09] Intelligent Living (10 January 2020)
Andrea D. Steffen
Our solar system is surrounded by an 89,000°F wall of plasma
bit.ly/supermars-4-09

Chapter 5

[5-01] Scientific American (25 February 2015) Robin Wylie
Giant asteroid collision may have radically transformed Mars
bit.ly/supermars-5-01

[5-02] Cosmos (2 May 2019) Richard A. Lovett
Earth hit by 17 meteors a day
bit.ly/supermars-5-02

[5-03] Space.com (29 April 2020) Chelsea Gohd
Death from above: Scientists find earliest evidence of person killed by meteorite
bit.ly/supermars-5-03

[5-04] Mysterious Universe (31 July 2019) Jocelyne LeBlanc
Surprising new theory suggests an ancient impact on Mars created a mega-tsunami
bit.ly/supermars-5-04

[5-05] Space.com (19 March 2018) Charles Q. Choi
Ancient oceans on Mars may have been older and shallower than thought
bit.ly/supermars-5-05

[5-06] Wikipedia (accessed 23 November 2020)
Tectonics of Mars
bit.ly/supermars-5-06

[5-07] RT (21 December 2017) James Moore and Jon Wade
How did Mars lose its oceans? Scientist may have cracked the mystery
on.rt.com/8vce

[5-08] Futurism (15 September 2020) Victor Tangermann
Acids may have destroyed evidence of life on Mars
bit.ly/supermars-5-08

[5-09] Scientific American (26 June 2008) JR Minkel
Pay dirt: Martian soil fit for Earthly life
bit.ly/supermars-5-09

[5-10] Space.com (12 December 2017) Tim Sharp
What is Mars made of? Composition of planet Mars
bit.ly/supermars-5-10

[5-11] Environmental Operating Solutions, Inc. (accessed 23 November 2020)
Perchlorate
bit.ly/supermars-5-11

[5-12] Futurism (27 August 2020) Jon Christian
Scientists: Bacteria could survive trip to Mars on outside of spacecraft
bit.ly/supermars-5-12

[5-13] Futurism (24 July 2020) Dan Robitzski
Oh great: space travel makes bacteria even deadlier
bit.ly/supermars-5-13

[5-14] Space.com (29 November 2017) Mike Wall
Bacteria 'from outer space' found on space station, cosmonaut says: report
bit.ly/supermars-5-14

[5-15] New Scientist (1 February 2012) Michael Marshall
First land plants plunged Earth into ice age
bit.ly/supermars-5-15

[5-16] New Scientist (31 May 2007) David Shiga
Lab study indicates Mars has a molten core
bit.ly/supermars-5-16

Chapter 6

[6-01] Futurism (28 February 2019) Kristin Houser
First evidence of "planet-wide groundwater system" on Mars found
bit.ly/supermars-6-01

[6-02] Phys.org (1 March 2019)
Markus Bauer, European Space Agency
First evidence of planet-wide groundwater system on Mars
bit.ly/supermars-6-02

[6-03] Interesting Engineering (15 February 2017)
Christopher McFadden
Could we actually create artificial gravity in space?
bit.ly/supermars-6-03

[6-04] AnandTech (6 August 2005)
various comments on message board
How fast does a space station have to spin to generate gravity?
bit.ly/supermars-6-04

[6-05] University of Washington (8 June 2015)
Peter Kelley
Atmospheric signs of volcanic activity could aid search for life
bit.ly/supermars-6-05

[6-06] Science Daily (2 March 2016)
Great tilt gave Mars a new face
bit.ly/supermars-6-06

[6-07] Celebrate Birth (16 April 2019)
Do your odd pregnancy cravings mean something?
bit.ly/supermars-6-07

[6-08] University of Washington
(accessed 24 November 2020) James E. Tillman
Mars: temperature overview
bit.ly/supermars-6-08

[6-09] Wikipedia (accessed 24 November 2020)
Diurnal temperature variation
bit.ly/supermars-6-09

[6-10] Stack Exchange (27 January 2017)
various comments on message board
Solar panels on Mars?
bit.ly/supermars-6-10

[6-11] Space.com (13 June 2013) Leonard David
Toxic Mars: astronauts must deal with perchlorate on the red planet
bit.ly/supermars-6-11

[6-12] USGS (accessed 24 November 2020)
What would happen if a "supervolcano" eruption occurred again at Yellowstone?
on.doi.gov/3pWDU2E

Chapter 7

[7-01] Wikipedia (accessed 24 November 2020)
Proxima Centauri
bit.ly/supermars-7-01

[7-02] Universe Today (2008) Fraser Cain
What are the different types of stars?
bit.ly/supermars-7-02

[7-02a] Enchanted Learning (accessed 7 December 2020)
Star Types
bit.ly/supermars-7-02a

[7-03] Science (2 April 2015) Ann Gibbons
How Europeans evolved white skin
bit.ly/supermars-7-03

Chapter 8

[8-01] Futurism (10 October 2019) Victor Tangermann
Former NASA scientist "convinced" we already found life on Mars
bit.ly/supermars-8-01

[8-02] Futurism (27 August 2020) Jon Christian
Scientists: bacteria could survive trip to Mars on outside of spacecraft
bit.ly/supermars-8-02

[8-03] Space.com (9 August 2019) Mike Wall
Life may be common in the Milky Way, thanks to comet swapping
bit.ly/supermars-8-03

[8-04] NASA JPL (accessed 24 November 2020) Ron Baalke
Mars meteorites
go.nasa.gov/3pTUOPn

[8-05] New Scientist (26 August 2020) Layal Liverpool
Radiation-resistant bacteria could survive journey from Earth to Mars
bit.ly/supermars-8-05

[8-06] EarthSky (8 August 2019) Paul Scott Anderson
A mega-tsunami on ancient Mars?
bit.ly/supermars-8-06

[8-07] New Scientist (24 June 2019) Donna Lu
Mars meteorite assault stopped 500 million years earlier than thought
bit.ly/supermars-8-07

[8-08] New Scientist (13 March 2020) Michael Marshall
Mars may once have had right conditions for RNA to develop into life
bit.ly/supermars-8-08

[8-09] Futurism (22 October 2020) Dan Robitzski
Researchers: claim of life molecules on Venus may have been an error
bit.ly/supermars-8-09

[8-10] Futurism (1 April 2019) Dan Robitzski
Astronomers found Mars' missing methane
bit.ly/supermars-8-10

[8-11] Science Alert (1 April 2019) Michelle Starr
It's official: we now have independent confirmation of methane on Mars
bit.ly/supermars-8-11

[8-12] BBC (28 July 2020)
Scientists revive 100 million-year-old microbes from the sea
bit.ly/supermars-8-12

[8-13] Mysterious Universe (2 May 2017) Brent Swancer
The digital universe: on living in a computer simulation
bit.ly/supermars-8-13

[8-14] Mysterious Universe (29 September 2019)
Jocelyne LeBlanc
Earliest life on Earth discovered in Australian desert
bit.ly/supermars-8-14

Chapter 9

[9-01] Discover (25 September 2018) Amber Jorgenson
Mars could have had underground life for millions of years
bit.ly/supermars-9-01

[9-02] The Washington Post (30 March 2017)
Joel Achenbach
How Mars lost its atmosphere, and why Earth didn't
wapo.st/3nSvwzu

[9-03] Wikipedia (accessed 24 November 2020)
Gale crater
bit.ly/supermars-9-03

[9-04] The New York Times (9 December 2013)
Kenneth Chang
Ancient Martian lake may have supported life
nyti.ms/39aMQvb

[9-05] Gizmodo (7 August 2019) George Dvorsky
NASA's Curiosity rover stumbles upon a strangely complicated Martian rock
bit.ly/supermars-9-05

[9-06] Nature (2 June 2005) Mark Peplow
Martian methane could come from rocks
go.nature.com/35WE8yJ

[9-07] Futurism (24 June 2019) Victor Tangermann
New NASA rover discovery hints at signs of life on Mars
bit.ly/supermars-9-07

[9-08] Science.com (27 March 2019) Mike Wall
Mars had big rivers for billions of years
bit.ly/supermars-9-08

[9-09] Futurism (28 March 2019) Victor Tangermann
New research: Mars used to be covered in huge rivers
bit.ly/supermars-9-09a

[9-10] RT (21 December 2018)
Water on Mars pictured: ESA shares incredible images of Martian ice crater
on.rt.com/9l0j

[9-11] New Scientist (29 March 2019) Leah Crane
Water on Mars is probably too cold and salty for life as we know it
bit.ly/supermars-9-11

[9-12] Independent (11 May 2020) Andrew Griffin
Alien life probably could not thrive in Martian water, researchers say
bit.ly/supermars-9-12

[9-13] Futurism (29 March 2019) Jon Christian
Researchers detect "deep groundwater" on Mars
bit.ly/supermars-9-13

[9-14] BBC (23 April 2019) Jonathan Amos
NASA's InSight lander 'detects first Marsquake'
bbc.in/3m0L0kj

[9-15] Futurism (7 July 2020) Victor Tangermann
Scientists: flying by Venus to get to Mars would be cheaper, faster
bit.ly/supermars-9-15

[9-16] Futurism (12 May 2020) Dan Robitzski
Scientists say they've found the perfect spot for a Mars colony
bit.ly/supermars-9-16

[9-17] Futurism (12 February 2020) Victor Tangermann
This group is collecting designs for a Martian city
bit.ly/supermars-9-17

[9-18] Futurism (11 June 2020) Victor Tangermann
This plan for a Martian city under a dome is breathtaking
bit.ly/supermars-9-18

[9-19] Futurism (13 August 2019) Victor Tangermann
Elon Musk: Mars city could cost up to $10 trillion
bit.ly/supermars-9-19

[9-20] Futurism (8 November 2019) Victor Tangermann
Elon Musk: I can build a Martian city with 1,000 starships
bit.ly/supermars-9-20

[9-21] Space.com (14 June 2017) Mike Wall
SpaceX's Mars colony plan: how Elon Musk plans to build a million-person Martian city
bit.ly/supermars-9-21

[9-22] Futurism (17 January 2020) Victor Tangermann
Elon Musk says he'll put a million people on Mars by 2050
bit.ly/supermars-9-22

[9-23] CNET (16 January 2020) Amanda Kooser
Elon Musk drops details for SpaceX Mars mega-colony
cnet.co/3kYLQwr

[9-24] Futurism (24 October 2019) Dan Robitzski
NASA's collaborating with Caterpillar on Moon mining machines
bit.ly/supermars-9-24

[9-25] Futurism (16 January 2020) Dan Robitzski
NASA wants to grow a Moon base out of mushrooms
bit.ly/supermars-9-25

[9-26] Futurism (17 September 2020) Victor Tangermann
Scientists: we could build Mars shelters out of insect polymers and Martian soil
bit.ly/supermars-9-26

[9-27] Futurism (4 September 2019) Dan Robitzski
Robots could build Martian settlements by imitating humans on Earth
bit.ly/supermars-9-27

[9-28] Futurism (5 May 2020) Dan Robitzski
These self-repairing lunar habitats could help settle the Moon
bit.ly/supermars-9-28

[9-29] Futurism (13 August 2020) Victor Tangermann
Researchers built a "gravity suit" to keep astronauts healthy
bit.ly/supermars-9-29

[9-30] Futurism (20 May 2019) Natalie Coleman
Evolutionary biologist: Mars colonists will mutate really fast
bit.ly/supermars-9-30

[9-31] Tech Times (21 May 2019) Naia Carlos
First humans on Mars will evolve into an entirely new species quickly, says evolutionary biologist
bit.ly/supermars-9-31

[9-32] Wikipedia (accessed 24 November 2020)
Terraforming of Mars
bit.ly/supermars-9-32

[9-33] Gizmodo (30 July 2019) George Dvorsky
Humans will never colonize Mars
bit.ly/supermars-9-33

[9-34] Futurism (25 November 2019) Dan Robitzski
Professor: terraforming a new home planet would be dangerous
bit.ly/supermars-9-34

[9-35] Extreme Tech (6 March 2017) Ryan Whitwam
NASA proposes building artificial magnetic field to restore Mars' atmosphere
bit.ly/supermars-9-35

[9-36] Stack Exchange (11 March 2018)
Various comments on message board
How far would the Mars L1 Lagrangian Point be from Mars?
bit.ly/supermars-9-36

[9-37] New Scientist (14 October 2020)
Jonathan O'Callaghan
The Moon had a magnetic field that helped protect Earth's atmosphere
bit.ly/supermars-9-37

[9-38] New Scientist (25 September 2020) Layal Liverpool
Radiation exposure on the Moon is nearly three times that on the ISS
bit.ly/supermars-9-38

[9-39] New Scientist (14 October 2020)
Jonathan O'Callaghan
The Moon had a magnetic field that helped protect Earth's atmosphere
bit.ly/supermars-9-37

[9-40] Gizmodo (23 May 2019) George Dvorsky
An astounding amount of water has been discovered beneath the Martian north pole
bit.ly/supermars-9-40

[9-41] Futurism (31 July 2018) Kristin Houser
Scientists say we can't terraform Mars. Elon Musk says we can.
bit.ly/supermars-9-41

[9-42] Fandom: Fallout Wiki (5 November 2009)
Various comments on message board
Why does everyone think 200 years would get rid of all the radiation?
bit.ly/supermars-9-42

[9-43] Mysterious Universe (19 August 2019)
Sequoyah Kennedy
Elon Musk still wants to nuke Mars
bit.ly/supermars-9-43

[9-44] Futurism (19 August 2019) Dan Robitzski
Sorry, Elon: terraforming Mars would take 3,500 nukes per day
bit.ly/supermars-9-44

[9-45] Science 2.0 (17 August 2019) Robert Walker
Nukes can never terraform Mars
bit.ly/supermars-9-45

[9-46] Futurism (21 August 2019) Victor Tangermann
Elon Musk: "Thousands of solar reflector satellites" could warm up Mars
bit.ly/supermars-9-46

[9-47] Next Big Future (16 August 2019) Brian Wang
Terraforming Mars in 50 years with large orbital mirrors, bacteria and factories
bit.ly/supermars-9-47

[9-48] Futurism (27 May 2020) Dan Robitzski
This video of Mars' leaking atmosphere could make Elon Musk cry
bit.ly/supermars-9-48

[9-49] Futurism (21 November 2019) Dan Robitzski
Unusual PhD thesis: let's use bacteria to colonize Mars
bit.ly/supermars-9-49

[9-50] New Scientist (27 November 2019) Gege Li
CO_2-guzzling bacteria made in the lab could help tackle climate change
bit.ly/supermars-9-50

[9-51] MarketWatch (19 November 2016) Jurica Dujmovic
Here's how we could warm up Mars and make it more habitable
on.mktw.net/3pW2mBa

[9-52] Futurism (30 July 2019) Dan Robitzski
Reality check: it would take thousands of years to colonize Mars
bit.ly/supermars-9-52

[9-53] NASA (17 November 2008) Kathryn Hansen
Water vapor confirmed as major player in climate change
go.nasa.gov/373hJiK

[9-54] Ars Technica (28 January 2015) Xaq Rzetelny
Where did Earth's nitrogen come from?
bit.ly/supermars-9-54

[9-55] Memorial University (2014) Steven M. Carr
The Meselson - Stahl experiment: proof of semi-conservative replication
bit.ly/supermars-9-55

[9-56] Science Alert (4 May 2020) Carly Cassella
Scientists have discovered fixed nitrogen in a Martian meteorite for the first time
bit.ly/supermars-9-56

[9-57] NPR (27 August 2020) Nell Greenfieldboyce
Water, water, everywhere – and now scientists know where it came from
n.pr/371wLpe

[9-58] Royal Belgian Institute for Space Aeronomy (accessed 24 November 2020)
Mars, atmosphere without ozone layer
bit.ly/supermars-9-58

[9-59] Tomatosphere/Let's Talk Science (accessed 24 November 2020)
Is there enough light on Mars to grow plants?
bit.ly/supermars-9-59

[9-60] Science Alert (9 March 2016) Fiona MacDonald
Tomatoes, peas, and 8 other crops have been grown in Mars-equivalent soil
bit.ly/supermars-9-60

[9-61] Eat Like A Martian (accessed 24 November 2020) Kevin Cannon
What will people eat on Mars?
bit.ly/supermars-9-61

[9-62] Futurism (27 September 2019) Natalie Coleman
Contaminating Mars with microbes could kickstart colonization
bit.ly/supermars-9-62

[9-63] Wikipedia (accessed 24 November 2020)
Interplanetary contamination
bit.ly/supermars-9-63

[9-64] Futurism (19 May 2020) Dan Robitzski
Japanese startup wants to replace astronauts with space robots
bit.ly/supermars-9-64

[9-65] Futurism (13 February 2020) Jon Christian
NASA: Mars astronauts will use lasers to communicate with Earth
bit.ly/supermars-9-65

[9-66] BBC Future (30 October 2019) Kelly Oakes
How long space voyages could mess with our minds
bbc.in/3l20d39

[9-67] Wikipedia (accessed 24 November 2020)
Human mission to Mars
bit.ly/supermars-9-67

[9-68] New Scientist (14 November 2013) Lisa Grossman
Mars probe to sniff atmosphere and scout safer landings
bit.ly/supermars-9-68

[9-69] Science Daily/Southwestern University (21 August 2019)
Spaceflight consistently affects the gut
bit.ly/supermars-9-69

[9-70] Scientific American (15 July 2019) Mike Wall
Silica blankets could make Mars habitable
bit.ly/supermars-9-70

[9-71] Futurism (7 November 2019) Dan Robitzski
This scientist wants to gene-hack hybrid humans to survive Mars
bit.ly/supermars-9-71

[9-72] Space.com (19 May 2020) Mike Wall
Colonizing Mars may require humanity to tweak its DNA
bit.ly/supermars-9-72

[9-73] Curiosmos (17 May 2020) Ivan Petricevic
Who says aliens on distant planets need oxygen to survive?
bit.ly/supermars-9-73

[9-74] Space.com (19 May 2020) Mike Wall
Colonizing Mars may require humanity to tweak its DNA
bit.ly/supermars-9-72

[9-75] Futurism (26 April 2019) Victor Tangermann
This Martian greenhouse just won a NASA award
bit.ly/supermars-9-75

[9-76] Futurism (25 September 2019) Kristin Houser
Here's how we could feed a million people on Mars
bit.ly/supermars-9-76

[9-77] Space.com (18 September 2019) Charles Q. Choi
How to feed a Mars colony of 1 million people
bit.ly/supermars-9-77

[9-78] Live Science (15 January 2013) Christopher Wanjek
Reality check: 5 risks of raw vegan diet
bit.ly/supermars-9-78

[9-79] Futurism (16 December 2019) Dan Robitzski
A former NASA astronaut is building a plasma-powered Mars rocket
bit.ly/supermars-9-79

[9-80] Mysterious Universe (23 November 2019) Jocelyne LeBlanc
Living on Mars could cause humans to develop dementia
bit.ly/supermars-9-80

[9-81] Space.com (6 February 2020) Chelsea Gohd
How to die on Mars
bit.ly/supermars-9-81

[9-82] New Scientist (19 June 2020) Leah Crane
Electric dust could be erasing signs of life from the surface of Mars
bit.ly/supermars-9-82

[9-83] Wikipedia (accessed 25 November 2020)
Terraforming of Venus
bit.ly/supermars-9-83

[9-84] Universe Today (5 June 2020) Evan Gough
Interstellar Oumuamua was a dark hydrogen iceberg
bit.ly/supermars-9-84

[9-85] Science 2.0 (15 December 2013) Robert Walker
Trouble with terraforming Mars
bit.ly/supermars-9-85

Chapter 10

[10-01] BBC Future (3 November 2020)
Nicholas R. Longrich
Did Neanderthals go to war with our ancestors?
bbc.in/36zoEJG

[10-02] New Scientist (16 October 2019)
Ruby Prosser Scully
Damping down brain cell activity may help us to live longer
bit.ly/supermars-10-02

[10-03] New Scientist (12 February 2020) Leah Crane
Mars may have formed 15 million years later than we thought
bit.ly/supermars-10-03

[10-04] Wikipedia (accessed 25 November 2020)
London Hammer
bit.ly/supermars-10-04

[10-05] MessageToEagle.com (11 June 2014)
Did ancient aliens leave a 300-million-year-old aluminum gear on Earth?
bit.ly/supermars-10-05

[10-06] BBC Future (3 November 2020)
Nicholas R. Longrich
Did Neanderthals go to war with our ancestors?
bbc.in/36zoEJG

[10-07] Wikipedia (accessed 25 November 2020)
Hueyatlaco
bit.ly/supermars-10-07

[10-08] BBC Future (3 November 2020)
Nicholas R. Longrich
Did Neanderthals go to war with our ancestors?
bbc.in/362oEJG

[10-09] Gizmodo (18 September 2020) George Dvorsky
120,000-year-old human footprints mark possible migration route through Arabian peninsula
bit.ly/supermars-10-09

[10-10] Australian Museum (27 May 2020) Fran Dorey
How have we changed since our species first appeared?
bit.ly/supermars-10-10a

[10-11] Fast Company (1 November 2016) Charlie Sorrell
Over the last 2,000 years, humans have evolved in some surprising ways
bit.ly/supermars-10-11

[10-12] ABC Australia (14 May 2019) Antony Funnell
It isn't obvious, but humans are still evolving. So what will the future hold?
ab.co/2HBd0vZ

[10-13] BBC Earth (accessed 25 November 2020)
Lucy Jones
What will humans look like in a million years?
bit.ly/supermars-10-13

[10-14] Scientific American (1 November 2012) Peter Ward
What may become of *Homo sapiens*
bit.ly/supermars-10-14

[10-15] Wikipedia (accessed 25 November 2020)
Future of Earth
bit.ly/supermars-10-15

[10-16] Futurism (9 September 2020) Dan Robitzski
Too much CO2 is killing trees, scientists say
bit.ly/supermars-10-16

[10-17] Wikipedia (accessed 25 November 2020)
Timeline of the far future
bit.ly/supermars-10-17

Chapter 11

[11-01] Disabled World (30 November 2017)
Average height to weight chart: babies to teenagers
bit.ly/supermars-11-01

[11-02] Rice University (17 January 2018) Scott Solomon
TEDx talk
bit.ly/supermars-11-02

[11-03] Futurism (14 November 2019) Victor Tangermann
NASA research: astronauts are getting clots, bizarre blood flow
bit.ly/supermars-11-03

[11-04] Futurism (15 November 2019) Dan Robitzski
New research: astronauts are experiencing cognitive decline
bit.ly/supermars-11-04

[11-05] Wikipedia (accessed 25 November 2020)
Mars Gravity Biosatellite
bit.ly/supermars-11-05

[11-06] Futurism (23 July 2019) Kristin Houser
Biologist: babies born in space might not be fully human
bit.ly/supermars-11-06

[11-07] NBC News (28 February 2017) David Freeman
Will Mars colonists evolve into this new kind of human?
https://nbcnews.to/3m6H9lP

[11-08] Business Insider (25 May 2018)
María Soledad González Romero and Jordan Bowman
What humans will look like on Mars
bit.ly/supermars-11-08

Chapter 12

[12-01] National Space Society
(accessed 25 November 2020)
NSS Roadmap to Space Settlement (3rd edition)
bit.ly/supermars-12-01

[12-02] The Washington Post (1 May 2019) Buzz Aldrin
Buzz Aldrin: It's time to focus on the great migration of humankind to Mars
wapo.st/3770flB

[12-03] NASA (17 May 2016)
Blaine Friedlander, University of Cornell
Giant red stars may heat frozen worlds into habitable planets
go.nasa.gov/362rBdg

[12-04] Wikipedia (accessed 25 November 2020)
Colonization of the outer Solar System
bit.ly/supermars-12-04

[12-05] Futurism (2 November 2020) Dan Robitzski
NASA: half the Sun-like planets in our galaxy have potentially habitable planets
bit.ly/supermars-12-05

[12-06] Wikipedia (accessed 25 November 2020)
Future of space exploration: limitations with deep space exploration
bit.ly/supermars-12-06

[12-07] Futurism (9 July 2020) Dan Robitzski
On an interstellar flight, language itself would evolve
bit.ly/supermars-12-07

[12-08] MIT News (26 November 2019) Janine Liberty
Designing humanity's future in space
bit.ly/supermars-12-08

[12-09] Futurism (8 July 2020) Victor Tangermann
Scientists ponder gory mayhem of zero-gravity surgery
bit.ly/supermars-12-09

[12-10] Futurism (18 November 2019) Victor Tangermann
Scientist: astronaut hibernation pods "actually not so crazy"
bit.ly/supermars-12-10

[12-11] Science Alert (10 April 2020) Michelle Starr
Homo Galacticus: how space will shape the humans of the future
bit.ly/supermars-12-11

[12-12] Wikipedia (accessed 25 November 2020)
O'Neill cylinder
bit.ly/supermars-12-12

[12-13] Ars Technica (10 May 2019) Eric Berger
Jeff Bezos unveils his sweeping vision for humanity's future in space
bit.ly/supermars-12-13

[12-14] Space.com (27 May 2014) Nola Taylor Redd
Future of space exploration could see humans on Mars, alien planets
bit.ly/supermars-12-14

[12-15] Futurism (24 June 2019) Victor Tangermann
Study: sperm banks in space could make colonizing Mars possible
bit.ly/supermars-12-15

Chapter 13

[13-01] Space.com (28 March 2018) Nola Taylor Redd
Red giant stars: facts, definition & the future of the Sun
bit.ly/supermars-13-01

[13-02] Gizmodo (16 September 2020) George Dvorsky
Astronomers discover first known planet to orbit a white dwarf star
bit.ly/supermars-13-02

[13-03] Sciencing (24 April 2017) John Papiewski
What kind of stars live the longest?
bit.ly/supermars-13-03

[13-04] Wikipedia (accessed 25 November 2020)
Habitability of red dwarf systems
bit.ly/supermars-13-04

[13-05] Scientific American (19 November 2012)
Caleb A. Scharf
The stars are beginning to go out...
bit.ly/supermars-13-05

[13-06] New Scientist (14 July 2020) Donna Lu
Earth-sized exoplanets around small stars may have intense hurricanes
bit.ly/supermars-13-06

[13-07] Futurism (7 October 2020) Dan Robitzski
Physicist: there were other universes before the Big Bang
bit.ly/supermars-13-07

15
About the Author

Dr. Ellis Silver is an American ecologist and environmentalist.

He now spends the majority of his time in Europe, the Middle East, and Indonesia – mostly on boats.

He is happy to answer sensible questions by email – on the rare occasions when he has a decent internet connection.

Email address: silver.ellis@gmail.com

Facebook page: facebook.com/HumansAreNotFromEarth

16
Index

Adams, Douglas, 152
aging, 113
air pressure, 52
air, composition of, 166
albedo, reducing, 246
Alpha Centauri, 29
antimatter, 373
asteroids, 155
   impacts, 38, 133
Atacama Desert, 26
atmosphere, 15
   detoxifying, 251
   insulating, 256
   reinstating, 243

back problems, 70
basalt, 39
behavior, natural, 112
biopolymers, 235
birth issues, 90
blue light, 50

body clock, 75
body hair, 53
bone density, 71
brain size, 73, 336
broadcasts, 10

carbon dioxide, 16
   human reaction to, 87
   releasing, 244
Chernobyl fungus, 14
children versus plants, 88
chronic illness, 27
climate change, 132
   deliberate, 245
clothes, 53
Cold War, 30
colonization of Mars
   equipment, 296
communications, 10, 289
contamination, 204
coronal mass ejection, 136

cravings, 68
crops on Mars, 259
cyborgs, 272

defense mechanisms, 84
Deinococcus radiodurans, 183, 210
depression, 107
destruction, 99, 144
diet, 68
differences between Earth and Mars, 176
disasters, man-made, 138
diurnal temperature variation, 183
DNA
  anomalies, 101
  scars, 43
domed habitats, 233
dust, 300
  storms, 79
    eradicating, 261
dynamo effect, 15, 128
dyskinesia, 50

Earth
  demise, 346
  evolution of life, 205
  moving, 348
Eden, 8, 187
  age, 129, 188
  atmosphere, 194
  climate, 196
  fauna, 201
  flora, 199
  geological activity, 193
  location, 187
  magnetic field, 193
  mineral composition, 197
  returning to, 350
  rotation and tilt, 192
  size and gravity, 189
  water, 198
embryos, frozen, 378
EmDrive, 276
encephalitis, 74, 283
environment
  collapse, 137
  disaster, 30
  response, 60
erosion, 40
evidence, 47, 49
evolution
  design flaws, 121
  political and societal, 366
  theory of, 130
extinction, 146
extremophiles, 165, 182

famine, 137
food, natural, 67
food production, 272
  farming, 297
future,
  in space, 369
  ultimate, 383

Gale crater, 26, 224
genome, modification, 43
geological activity, 158
geology, 72
    planets compared, 173
glaciers, 12
grandparenting, 115
gravity
    artificial, 177, 180
    effects of, 267
Great Oxygenation Event, 16, 167, 205, 214

habitability, 11
habitats, 14
halocarbon, 248
hay fever, 56
heads, enlarged
    (see *encephalitis*)
healing powers, 111
height, 334
Hellas Planitia, 13
help, extraterrestrial, 31
hibernation, 59
hominins, native, 19
human evolution
    future, 342
    on Earth, 334,
    timeline, revised, 319
human habitation, 230
human reproduction, 364
humans on Earth, 313
hydroponics, 272

illness, chronic, 61
impairment, 42
impulse drive, 277
insects, 11
Instituto Pio XII, 52
Intelligent Design, 131
interbreeding, 43
International Space Station, 13
Into Africa theory, 44

KBC void, 9
Krakatoa, 185

L1 Lagrangian Point, 240
Labeled Release, 203
lava tubes, 232
Levin, Gilbert V., 203
life, 37
    detection, 209
    evolution on Earth, 211
    evolution on Mars, 208
    interplanetary, 204
life on Mars, 203
    location, 215
    timeline, 217
living on Mars, 280

magnetic field, 15
    detecting, 81
    Venus, 128
magnetosphere
    loss of, 127
    reinstating, 240-241

Marinova, Margarita, 249
Mars
   acclimatizing, 365
   birth issues, 361
   colonizing, 262
   future demise of, 370
   gravity, effects of, 355
   habitats, 264
   history of, 153, 222
   human evolution, 361
   human future, 355, 367
   seasons, 156
   similarity to Earth, 174
   water, 225
marsquakes, 79, 229
mass extinction, 34
M-dwarf stars, 387
memory, erasing, 42
menopause, 115
meteors, 155
methane, 210, 223
microbiome, 266
mirrors, solar, 183, 244
missing link, 106
moon, loss of, 134
morning sickness, 94
motivation, 372
Musk, Elon, 234, 244, 279

NASA, 203
natural disasters, 127
   sensing, 78
Neanderthals, 27

neophobia, 88
nitrogen, 252
northern ocean, 36
nuclear bombs, 244

O'Neill cylinders, 379
Olympus Mons, 158
orbit, 180
out of place, feeling, 47
overpopulation, 85, 143
ovulation, concealed, 96
oxygen, 16
ozone, 35
   restoring, 257

pandemics, 137, 140
panspermia, 23, 25, 210
   human, 381
parasites, 69
penis bone (baculum), 104
perchlorate, 162
perfluorocarbon, 249
phosphine, 209
photodecomposition, 34
photonic propulsion, 277
Plan B, 18
planets, finding, 374
plants, edible, 47
pressure, atmospheric, 36
Prison Planet Theory, 149
Proxima Centauri, 187, 371
pseudoscience, 9
Pye, Lloyd, 102

quantum entanglement, 289

racial diversity, 337
radiation, 163
   human response, 288
   shielding, 13-14, 232
rape theory, 315
red dwarf stars, 386
red giant stars, 135, 384
rhesus negative blood, 94
rivers on Mars, 12
Roswell Incident, 118

self-destruction, 107
selfishness and greed, 98
self-sufficiency, 270
semi-aquatic features, 102
sense of direction, 80
skills, savant, 116
skull, shape of, 103
soil, 162
   detoxifying, 258
solar flares
   (see *coronal mass ejection*)
speciation, 361
sport on Mars, 295
Sun
   damage caused by, 49
   red giant phase, 23
super-intelligence, 100
super-Mars, 12
surgery in space, 375
sweet tooth, 89

temperature, maintaining, 255
terraforming Mars, 221, 238
   objections, 238
terraforming Venus, 306
terrorism, 142
thriving, 77
travel, space, 276
tsunami on Mars, 157, 207

ultimate fate, 389
Ultraviolent Warrior Theory, 147
underground habitats, 231
Uranus, 181, 370
UVB radiation, 54

Venus, 33, 306
violence, 19, 96
vitamin D, 50

war
   Earth–Mars, 97
   human–Neanderthal, 316
warp drive, 29
water, 36, 160
   drinking, 65
   restoring, 256
water vapor, 252
window of opportunity, 205
wormholes, 29, 289
worshipping deities, 110

zero gravity, life in, 375

Made in the USA
Middletown, DE
14 August 2023

36739036R00245